Access to B-ISDN via PONs

Access to B-ISDN via PONs

ATM Communication in Practice

Edited by

Ulrich Killat
Technische Universität Hamburg-Harburg
Germany

⊛WILEY 田TEUBNER

A Partnership between John Wiley & Sons and B. G. Teubner Publishers

Chichester · New York · Brisbane · Toronto · Singapore · Stuttgart · Leipzig

Copyright © 1996 jointly by John Wiley & Sons Ltd. and B.G. Teubner

John Wiley & Sons Ltd.   B.G. Teubner
Baffins Lane, Chichester   Industriestraße 15
Chichester   70565 Stuttgart (Vaihingen)
West Sussex PO19 IUD   Postfach 80 10 69
England   70510 Stuttgart
                                          Germany

National Chichester (01243) 779777   National Stuttgart (0711) 789010
International +44 1243 779777   International +49 711 789010

Other Wiley Editorial Offices

John Wiley & Sons, Inc., 605 Third Avenue,
New York, NY 10158-0012, USA

Brisbane • Toronto • Singapore •

Other Teubner Editorial Offices

B.G. Teubner, Verlagsgesellschaft mbH, Johannisgasse 16,
D-04103 Leipzig, Germany

Die Deutsche Bibliotheck - CIP-Einheitsaufnahme

Killat, Ulrich:
Access to B-ISDN via PONs : ATM communication in practice
/ ed. by U. Killat - Stuttgart; Leipzig; Teubner; Chichester;
New York; Brisbane; Toronto; Singapore : Wiley, 1996
    ISBN 3 519 06449 9 (Teubner)
    ISBN 0 471 95877 8 (Wiley)
WG : 37   DBN 94.686635.X   96.02.1
8057   nn

Library of Congress Cataloging-in-Publication Data

Access to B-ISDN via passive optical networks / edited by U. Killat.
        p.   cm.
    Work carried out under the auspices of the RACE Programme.
    ISBN 0471 95877 8 (alk. paper)
    1. Optical communications.   2. Integrated services digital
networks.   3. Broadband communication systems.   4. RACE (Progam)
I. Killat, U.
TK5103.59.A22   1996
621.382'75 — dc20
                                                    95-20542
                                                    CIP

British Library Cataloguing in Publication Data

A catalogue record for this book is available from the British Library

ISBN Wiley 0 471 95877 8
ISBN Teubner 3 519 06449 9

Typeset in 10/12pt Palatino by Laser Words, Madras
Printed and bound in Great Britain by Bookcraft (Bath) Ltd.

This book is printed on acid-free paper responsibly manufactured from sustainable fore-
station, for which at least two trees are planted for each one used for paper production.

Contents

Foreword

It is with great pleasure that I respond to the kind invitation of the BAF project to contribute prefatory remarks to this account of their work, carried out under the auspices of the RACE Programme (Research and Development in Advanced Communications in Europe).

The objective of the RACE Programme was to support the introduction of Integrated Broadband Communications in the European Union. An important part of this overall objective was served by the BAF project, which has aimed to produce a cost-effective access facility for broadband networks, especially for residential and small business customers. As this book relates, in order to do so the project consortium merged two advanced communications technologies, ATM and PON, with contributions from many other disciplines, to create a demonstrator which has been subject to extensive trials and testing which have been fruitful both in contributions to international standards and in development work to improve further future generations of the system. This book forms an important reference source through the experience gained in this unique experiment in advanced telecommunications.

Another significant feature of the project should not be overlooked however. When the European Commission first began looking at supporting industrial research in Europe in the mid-1980s, it was clear that we had important strengths in telecommunications. It was equally clear that a revolution was on the way — the digital revolution — and European pre-eminence in the field was not guaranteed for ever. The trend of convergence with the information technology industry (dominated by the Americans) and the microchip revolution which permits the mass-production of cheap, reliable high-tech equipment (at which the Japanese excel) presented the threat which the European industry faced then, and still faces today.

But Europe has great intellectual, scientific and industrial resources, well capable of meeting any challenge if effectively mobilised. By encouraging co-operative research in advanced communications, the RACE Programme has played its part in mobilising these resources. This is well shown by the BAF project, which has involved researchers from seven European countries working together to a common goal. It exploited academic expertise to be found at the National Technical University Athens, the Polytechnic University of Catalonia, Queen Mary and Westfield College UK, the Technical University of Hamburg-Harburg, the University of Nijmegen, the Polytechnic University of Lausanne, and the University of Stuttgart. Telecommunications researchers from the national institutes in Italy, Germany, the Netherlands and Spain supplied their knowledge and facilities. Four major and one smaller equipment manufacturer participated. All have contributed to the success of the BAF project, and through that to the success of the RACE Programme itself.

Dr Roland P O Hüber
Director DG XIII/B
European Commission

Preface

Engineers are familiar with hybrid systems. During their education they are usually confronted with clear cut principles leading to optimal solutions. The systems they build during their industrial careers however, are often hybrid. This results from a sort of compromise aimed at collecting all the strong points of different solutions while attempting to avoid their individual drawbacks.

This book is also a hybrid. It aims at meeting two goals and at the same time tries to reconcile the resulting conflicts stemming from different styles and ways of presentation. In the first two sections the book gives short tutorials on the topics of asynchronous transfer mode (ATM) and passive optical networks (PONs). Section 3 then describes the various aspects of developing an ATM-based PON. Sections 1 and 2 roughly represent the knowledge of the members of the team who started in January 1992 to build a system for which no forerunners are described in today's textbooks. The team was formed from a consortium of telecom companies, network operators and universities, and committed itself to building a broadband access network as a contribution to the RACE II project. The work was done under the name "Broadband Access Facilities (BAF)" and ended in June 1995.

The tasks of BAF were:

— to develop general concepts of what a broadband access network should look like
— to define a specific instance of the proposed architecture, and
— to specify, design, develop, test and demonstrate a prototype of the proposed ATM-based PON.

During the development phase the project was challenged by attempts to trade clear concepts for "quick and dirty solutions" which would perhaps have simplified and speeded up the implementation. It was a strength of the project and consequently of this book that the top-down approach was rigorously followed. Deviations from the generalised architecture are clearly indicated in the text and the reader can verify that they are only marginal. This book thus vividly illustrates how new concepts and new systems are created based on existing ones. The reader will also notice that the way to proceed need not be straightforward. The reason is that pushing certain system parameters to their limits may cause qualitative as well as the expected quantitative changes.

It may have been this effect which lead one of our internal reviewers to observe that Section 3 gives the impression of a series of almost unsolvable problems (for which solutions were eventually found). In a sense this observation is true. However, what is really remarkable is that, seen from the perspective of the completed system, the solutions to the admittedly non-trivial problems turned out to be not very difficult to realise and in most cases represented the effort of only some few square millimetres of silicon.

The different chapters of the book address by necessity a very broad scope of issues. Metastability in electronic circuitry, traffic theory and encryption are topics which do not seem at first glance to have a common denominator. Nevertheless, while avoiding becoming a patchwork, this book thoroughly addresses all the problems that need consideration in the design of a new communication system. There is a continuous thread from the foundations of Sections 1 and 2 to the experiments performed on the running prototype. Therefore the purpose of the book is twofold. It not only describes a piece of modern communication technology, but also illustrates the methodology of how problems are decomposed and solved making use of different engineering disciplines.

ACKNOWLEDGMENTS

I would like to take this opportunity to thank all members of the BAF team who directly or indirectly contributed to this book. The colleagues not appearing as authors for the chapters of this book but providing input for its contents are:

Laurenz Altwegg	N D Kalogeropoulos	Willem Romijn
Daniel Beeler	Amador Martin	Ronald Sennema
Geert-Jan Boecker	Anthony van Oyen	G I Stassinopoulos
Javier Escamilla	E N Protonotarios	Kerstin Uhde
Steve Glick	Alberto Profumo	Rob Verbeek

In addition, thanks go to Helmut Walle for his assistance in compiling and proofreading the chapters of the book. Finally, I want to express my thanks to Ilona Düring and Mathilde Winzer for making numerous corrections, additions and alterations without a word of criticism.

Ulrich Killat

List of Contributors

J Angelopoulos
NTUA National Technical University of Athens,
Athens

Th Apel
MAZ Mikroelektronik Anwendungszentrum Hamburg,
Hamburg

L d'Ascoli
ITALTEL,
Milano

Ch Blondia
University of Nijmegen,
Nijmegen

H Boekhorst
AT&T Network Systems Nederland,
Huizen

G Boukis
NTUA National Technical University of Athens,
Athens

J van Breemen
AT&T Network Systems Nederland,
Huizen

U Briem
University of Stuttgart,
Stuttgart

I Carretero
Telefónica I+D,
Madrid

O Casals
Polytechnic University of Catalonia,
Barcelona

J Charzinski
University of Stuttgart,
Stuttgart

L Cuthbert
Queen Mary and Westfield College,
London

M Dirksen
PTT Research,
Leidschendam

P Cr Fr Fonseca
DCID/Centro de Estudos de Telecomunicacoes

J Garcia
Polytechnic University of Catalonia,
Barcelona

U Heister
TUHH Technische Universität Hamburg-Harburg,
Hamburg

PH van Heijningen
AT&T Network Systems Nederland,
Huizen

M Kennedy
AT&T Network Systems Nederland,
Huizen

U Killat
TUHH Technische Universität
Hamburg-Harburg,
Hamburg

B Miah
Queen Mary and Westfield College,
London

Th Mosch
AT&T Network Systems Nederland,
Huizen

F Panken
University of Nijmegen,
Nijmegen

J Sanchez Cifuentes
Telefónica I+D,
Madrid

L Sara
Queen Mary and Westfield College,
London

P Schaafsma
AT&T Network Systems Nederland,
Huizen

R Slosiar
Ecole Polytechnique Fédérale de Lausanne,
Lausanne

P Solina
CSELT,
Torino

T Toniatti
ITALTEL,
Milano

S M Topliss
GPT Ltd.,
Coventry

M Valvo
CSELT,
Torino

I Venieris
NTUA National Technical University of
Athens,
Athens

H Walle
TUHH Technische Universität
Hamburg-Harburg,
Hamburg

R Widera
DBP Telekom FTZ,
Darmstadt

C-X Zhang
MAZ Mikroelektronik
Anwendungszentrum Hamburg,
Hamburg

List of Acronyms and Abbreviations

2B+D   Basic ISDN Access
AAL   ATM Adaptation Layer
ACSE   Association Control Service Element
A/D Converter   Analogue-to-Digital Converter
AFA   Active Fibre Amplifier
AM   Amplitude Modulation
AN   Access Network
ANE   Access Network Element
APD   Avalanche PhotoDiode
ASIC   Application Specific Integrated Circuit
ATM   Asynchronous Transfer Mode
AU-4   Administrative Unit 4 (SDH)
A-VPN   ATM-Virtual Private Network
AXC   ATM Cross-Connect
BAF   Broadband Access Facilities. Prefix indicating that the following notion is seen as developed by the BAF project for the BAF system.
BAF cell Each BAF cell consists of a three-octet preamble plus an ATM cell. The ATM cell format according to ITU-T I.361 recommendation at NNI.
BAF project The Broadband Access Facilities project supported by the EU under contract RACE 2024.
BAF system The ATM-PON system developed in the BAF project.
BECN   Backward Explicit Congestion Notification
BER   Bit Error Rate
BES   Best Effort Services
BG   Burst Generator
B-ISDN   Broadband Integrated Services Digital Network
Bit Alignment Upstream system synchronisation function at the OLT side, which regulates the phase of the data received from the different ONUs to the OLT system clock.
BK450 Standard for the delivery of TV signals in a 450 MHz band.
BK860 Standard for the delivery of TV signals in a 860 MHz band.
B-LT   Broadband Line Termination

B-NT1   Broadband Network Termination 1

B-NT2   Broadband Network Termination 2

BMR   Burst Mode Receiver

BMT   Burst Mode Transmitter

BP   Burst Processor

BPON   Broadband PON

BPP   Belt Permit Programming Protocol. A medium access protocol specifically developed for the BAF system.

BRAN   BRAnching Network

BSS   Bit and Slot Synchroniser

BTS   Bit Transport System

CAC   Connection Admission Control

CAD   Computer Aided Design

CAM   Computer Aided Manufacturing

CATV   Cable Television (also Common Antenna Television)

CBR   Constant Bit Rate

CC&M   Connection Control and Management

CCR   Commitment Concurrency and Recovery

CDMA   Code Division Multiple Access

CDV   Cell Delay Variation

CFS   Common Functional Specification

CLS   Connectionless Server

CMIP   Common Management Information Protocol

CMISE   Common Management Information Service Element

CMOS   Complementary Metal Oxide Semiconductor

CNR   Consiglio Nazionale delle Ricerche (National Research Council)

CON   Concentrator

CPE   Customer Premises Equipment

CPN   Customer Premises Network

CR   Coarse Ranging. Part of the BAF specific ranging procedure.

CRC   Cyclic Redundancy Code

CRG   Coarse Ranging Generator

CSCW   Computer Supported Cooperative Work

CSELT   Centro Studi e Laboratori Telecomunicazioni

CSLB   Continuous-State Leaky Bucket Algorithm

CTRL   ConTRoL

DAA   Dynamic Allocation Algorithm. A medium access protocol specifically developed for the BAF system.

DAP   Directory Access Protocol

DCN   Data Communication Network

DFR   Dynamic Fine Ranging. Part of the BAF specific ranging procedure.

DFSM   Dispersion Flattened Single Mode (fibre)

DFT   Discrete Fourier Transformation

DLL   Delay-Lock Loop

Downstream   From the core network to the customer premises network.

DQDB   Distributed Queue Dual Bus

DSP   Directory System Protocol

DSSM   Dispersion Shifted Single Mode (fibre)

DTP   Distributed Transaction Processing

DXC   Digital Cross-Connect

E/O   Electrical/Optical (conversion)

EDFA   Erbium Doped Fibre Amplifier

EDI   Electronic Data Interchange

Ethernet   Standard protocol for LANs (IEEE 802.3); used in the BAF project as a base for a management interface.

ETSI   European Telecommunications Standards Institute

EU   European Union

FDDI   Fibre Distributed Data Interface

FET   Field Effect Transistor

FG   Frame Generator

FOAM   Fast OAM. OAM commands that, due to efficiency reasons, are located in the preamble rather than in the payload of dedicated cells.

FP   Fabry–Perot (laser)

FP   Frame Processor

FTAM   File Transfer Access and Management

FTTB   Fibre-To-The-Building

FTTC   Fibre-To-The-Curb. An option in which the optical fibres from the OLT terminate at ONUs in the street. From the ONU, copper drops or some other medium are used to connect to several customers.

FTTH   Fibre-To-The-Home. An option in which the optical fibres from the OLT terminate at the customers premises at the ONUs.

GAP Same as guard time.

GBW   Guaranteed Bandwidth

GCRA   Generic Cell Rate Algorithm

GFC   Generic Flow Control

Global-FIFO Protocol A medium access protocol specifically developed for the BAF system.

Guard Time Period of time during which no ONU is allowed to transmit any data in the upstream direction; it has a nominal duration of 8 bits; the duration can actually change because of the finite accuracy of the DFR.

GUI   Graphical User Interface

HDTV   High-Definition TeleVision

HEC   Header Error Control

I/O   Input/Output

IC   Integrated Circuit

IEEE   Institute of Electrical and Electronics Engineers

IM-DD   Intensity Modulation-Direct Detection

IN   Intelligent Network

IP   Internet Protocol

ISDN   Integrated Services Digital Network

ITU   International Telecommunications Union

ITU-T Telecommunications standardisation sector of the International Telecommunications Union.

JPEG Joint Picture Expert Group

LAIP Local Exchange Access Network Interaction Protocol. A protocol running over the V interface to support exchange of information between the Access Network and the Local Exchange with the Local Exchange being the master.

LAN Local Area Network

LB Leaky Bucket

LCRF Local Related Control Functions

LD Laser Diode

LED Light-Emitting Diode

LEX Local Exchange. Core network element close to the access network.

LFSR Linear Feedback Shift Register

LPE LAIP Protocol Entity

LT Line Termination

MAC end-point Each MAC end-point is allowed to transmit upstream only when it receives a proper permit; there is one MAC end-point for each ONU (for transportation of OAM information), and one for each Tb and Tnb interface.

MAC Medium Access Control

MACO Message Authentication Code

Mc FRED Protocol A medium access protocol specifically developed for the BAF system.

MIB Management Information Base

MMPC Master MAC Protocol Controller

MPEG Moving Picture Expert Group

MTS Master Timing Source

MU Multiplexed Unit. Group of T-interfaces defined for the BPP Protocol.

MUX Multiplexer

NE Network Element

NNI Network–Network Interface (Network–Node Interface)

Normal Slot A normal slot is an upstream BAF cell.

NPC Network Parameter Control

NT Network Termination

NT1 Network Termination 1

NT2 Network Termination 2

O/E Optical/Electrical (conversion)

OAM Operation and Maintenance

ODN Optical Distribution Network

OLI Optical Line Inlet

OLO Optical Line Outlet

OLT Optical Line Termination. Network element system between the passive optical network and the LEX.

ONP Open Network Provisioning

ONU   Optical Network Unit. Network element unit between the passive optical network and the CPN.

ONU-C   Optical Network Unit-Curb. Network element unit between the passive optical network and the CPN in FTTC configurations, supporting up to eight Tb interfaces.

ONU-H   Optical Network Unit-Home. Network element unit between the passive optical network and the CPN in FTTH configurations, supporting only one Tb interface.

OPAL   Optische Anschlussleitung (Optical Access Line)

OS   Operating System. It incorporates the high-level management functions.

P&S   Privacy and Security

P/S   Parallel-to-Serial converter

Payload ATM layer information field of ATM cells, formed by 48 octets.

PCI   Protocol Control Information

PCM   Pulse Code Modulation

PDH   Plesiochronous Digital Hierarchy

PDU   Protocol Data Unit

PEAN   Pan-European ATM Network

Permit A PCI included in the downstream preamble. When a Tb interface queue receives a permit with its own code, it is allowed to send an upstream BAF cell.

P(I)N   Positive material (Intrinsic layer) Negative material

PLL   Phase-Lock Loops

PNO   Public Network Operator

PON   Passive Optical Network

POTS   Plain Old Telephone Service

Preamble The preamble defined in the BAF system consists of three octets, which are attached to the ATM cells crossing the U interface.

PRG   Pseudo-Random Generator

PRS   Pseudo-Random Sequence. Sequence generated by a PRG.

PSTN   Public Switched Telephone Network

PVC   Permanent Virtual Channel

PVP   Permanent Virtual Path

PWD   Pulse Width Distortion

QoS   Quality of Service

R&D   Research & Development

RAM   Random Access Memory

RAU   Request Access Unit. Special time slot in the BPP protocol.

RB   Request Block. Part of the BAF specific upstream frame structure; each RB consists of nine consecutive requests plus two stuffing octets.

RDA   Remote Database Access

RECIBA   Red Experimental de Comunicaciones Integradas de Banda Ancha

Request A six-octet PCI sent in the upstream direction to request a permit.

ROSE   Remote Operations Service Element

RTSE   Remote Transaction Service Element

Rx   Receiver

SCMA   SubCarrier Multiple Access
SDH   Synchronous Digital Hierarchy
SFR   Static Fine Ranging. Part of the BAF specific ranging procedure.
SMF   Service Management Function
SMPC   Slave MAC Protocol Controller
SNMP   Simple Management Network Protocol
SNR   Signal-to-Noise Ratio
SOA   Semiconductor Optical Amplifier
SOC   Slow OAM Cell. An OAM cell which contains messages interchanged between the OS and the ONUs across the PON.
SoD   Service on Demand
SPC   Service Provider Centre
SSM   Standard Single Mode (fibre)
STM   Synchronous Transfer Mode
STM-1   Synchronous Transport Module 1
STM-4   Synchronous Transport Module 4
STM-4c   Synchronous Transport Module 4 concatenated
SVC   Switched Virtual Channel
TAXI   Transparent Asynchronous Xmitter–receiver Interface
Tb   Broadband T interface.
TCP   Transmission Control Protocol
TDM   Time Division Multiplexing
TDMA   Time Division Multiple Access
TE   Terminal Equipment
Time Slot Transmission time of a BAF cell.
TMN   Telecommunications Management Network
Tnb   Narrowband T interface.
TPON   Telephony PON
Tx   Transmitter
UDP   User Datagram Protocol
UNI   User–Network Interface
UPC   Usage Parameter Control
Upstream From the CPN to the core network.
VBR   Variable Bit Rate
VC   Virtual Channel
VCC   Virtual Channel Connection
VCI   Virtual Channel Identifier
VoD   Video on Demand
VP   Virtual Path
VPC   Virtual Path Connection
VPCC   Virtual Path Cross Connect
VPI   Virtual Path Identifier
VPRN   Virtual Path Reference Number
VT   Virtual Terminal
WDM   Wavelength Division Multiplexing
WDMA   Wavelength Division Multiple Access

Section 1

THE ATM TECHNOLOGY

1.1 INTRODUCTION TO ATM-BASED B-ISDN

The public switched telephone network (PSTN) is the most widely used network in the world. It provides connectivity between more than half a billion telephone sets and this is still increasing. In the 1970s ideas came up

- to base this network on digital transport and switching technology
- to introduce an out-of-band signalling capability to facilitate the integration of voice and data services in the same network.

These objectives were further elaborated in the following decade. In 1984, the CCITT adopted a series of recommendations dealing with the Integrated Services Digital Network (ISDN). According to the CCITT, "an ISDN is a network, that provides end-to-end digital connectivity to support a wide range of services including voice and non-voice services, to which users have access by a limited set of standard multipurpose user–network interfaces". [1] The so-called basic access interface provided the user with two 64 kbit/s channels (B-channels) and a signalling channel of 16 kbit/s (D-channel). Of course such a proposal could not be the final step. End-users began to understand the competitive advantage of their networks and demanded solutions to their requirements in terms of computer communication and advanced video services. When thinking about a further evolution of ISDN into Broadband ISDN (B-ISDN) the first idea was to present a somewhat more sophisticated TDM-structure at the user–network interface including broadband channels in addition to those of the basic access interface. The approach failed because people were not able to agree on a set of channel types and a suitable multiplex structure. The inability to agree on the required set of parameters has to be considered as a failure of the concept itself. The type and number of channels offered to the subscriber need to be derived from a model of customer behaviour. While these models do exist for telephone traffic, they are hard to predict for forthcoming applications. Therefore, the key requirement for a future network has to be flexibility.

This flexibility can be expressed from a user perspective or from a network perspective.

- The user can dynamically create and use an (almost) arbitrary number of channels of (almost) arbitrary bandwidth.
- The network knows neither the syntax nor the semantics of the data it transports.

This decoupling of network and services should result in a manageable planning process requiring no detailed knowledge of services, their

Access to B-ISDN via Passive Optical Networks. Edited by U. Killat
© 1996 John Wiley & Sons Ltd

bandwidth demand or utilisation. With these ideas in mind, the CCITT has issued a new series of recommendations [2,3] which describe the asynchronous transfer mode (ATM) as "the target solution for B-ISDN". Initially, the fixed packet length multiplexing and switching technique, later called ATM, was driven by research and development in high-speed switching fabrics. [4,5] One of the decisive factors for making ATM a viable solution in almost every field of communication today was that it has been defined totally independent of an underlying transmission technology. Consequently, industries devoted to the local area network (LAN) market rather than to the public networks have taken up the challenge and have started marketing ATM-LANs. For the support of ATM in the private environment an organisation with several hundred member companies, the ATM Forum, has been created which issues recommendations [6,7] which are do not contradict CCITT (nowadays: ITU-T) standards, but which pave the way for a more rapid resolution of open issues. Many people consider the ATM Forum to be **the** driving force in ATM technology.

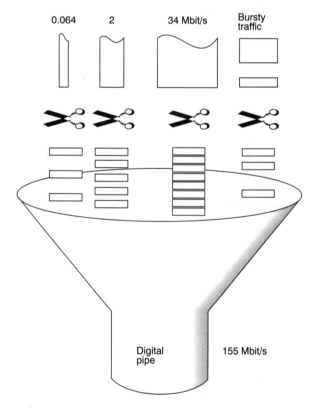

Figure 1.1.1
Visualisation of ATM principles

To understand the main principles of ATM, Figure 1.1.1 which has been widely used in conferences and papers on ATM is a good starting point. The figure visualises two basic concepts.

- Every type of information issued by a source is segmented into packets of equal length, which are referred to as "cells". After the segmentation process the frequency of cell arrivals is indicative of the instantaneous bandwidth demand of the source.

- Information from different sources is multiplexed on a single line by means of a statistical multiplexer.

The first concept is straightforward and easy to understand. A similar remark does not hold for the second concept. The stochastic processes describing the behaviour of ATM nodes, which can be thought of as being composed of multiplexer building blocks, are not easy to understand. Consequently estimate values and higher moments of the corresponding stochastic variables are not easily obtained but are badly needed in order to properly manage the resources of an ATM network. These problems will be discussed more intensively in Chapter 1.5. The description of the ATM technology presented in this book will follow the layered approach of system architecture which has been adopted from the work on ISDN and which is referred to as the B-ISDN protocol reference model (Figure 1.1.2).

A more refined description of layers and the functions contained therein is given in Figure 1.1.3, the details of which will be discussed in the chapters to follow. However, before doing so, there is a need to clarify the notion of a "cell" because it will be used for the description of both the ATM layer and the physical layer, but with a slightly different flavour.

A cell has already been introduced as a packet of fixed length as shown in Figure 1.1.4. The header of the cell contains a label for identification purposes and is protected by a cyclic redundancy check (CRC). In the physical layer the following types of cells are distinguished:

- valid cells (correct CRC)
- invalid cells (CRC failure)
- idle cells.

Idle cells represent a cell pattern which is used for stuffing purposes in order to adapt the user traffic to the capacity of the transmission system used. Only valid cells are passed to the ATM layer.

Before scrutinising the different layers of the reference model, another less abstract way of system decomposition known as "reference configuration" will be introduced. The reference configuration identifies functional groups and the reference points between them. The

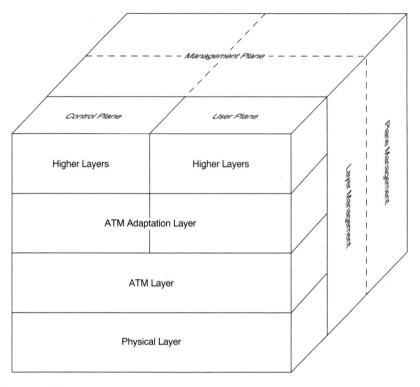

Figure 1.1.2
B-ISDN protocol reference model

ATM Adaptation Layer		CS	Convergence
		SAR	Segmentation and reassembly
ATM Layer		ATM	Generic flow control Cell header generation / extraction Cell VPI/VCI translation Cell multiplex and demultiplex
Physical Layer	Transmission Convergence Sublayer	TC	Cell rate decoupling HEC generation / verification cell delineation Transmission frame adaption Transmission frame generation / recovery
	Physical Medium Sublayer	PM	Bit timing Physical medium

Figure 1.1.3
Functions of the layers of the B-ISDN protocol reference model

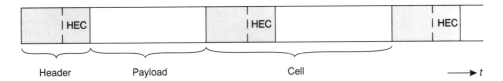

Figure 1.1.4
Concept of a cell

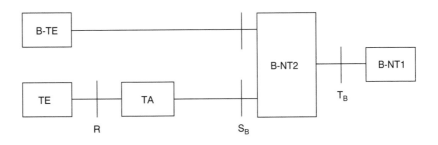

TE : Terminal equipment   R, S$_B$, T$_B$: Reference points
TA : Terminal adapter
NT : Network termination

Figure 1.1.5
B-ISDN user–network interface reference configuration

reference points identify where physical interfaces may occur between subsystems of a real network. An example for a B-ISDN user–network interface reference configuration is shown in Figure 1.1.5.

The generic term B-NT2 covers a couple of different actual system configurations: one extreme is the "null B-NT2" describing the direct connection of the terminal B-TE to the network termination B-NT1; in the other extreme the B-NT2 represents a customer premises network of LAN or PABX type.

The chapters of Section 1 will cover the layered architecture presented in Figure 1.1.2. Chapters 1.2 to 1.4 will cover the physical layer, the ATM layer and the ATM adaptation layer. Chapters 1.5 and 1.6 will address the problems that have to do with the non-deterministic nature of multiplexing and the resulting consequences for resource management. Finally, Chapter 1.7 will deal with the scenarios in which ATM technology is most likely to evolve in the future.

REFERENCES

[1]   CCITT recommendation I.110 *Preamble and General Structure of the I-Series Recommendations for the ISDN,* 1988
[2]   ITU-T (CCITT) recommendations

I.150 *B-ISDN ATM Functional Characteristics*, 1993
I.211 *B-ISDN Service Aspects*, 1993
I.311 *B-ISDN General Network Aspects*, 1993
I.327 *B-ISDN Functional Architecture*, 1993
I.356 *B-ISDN Layer Cell Transfer Performance*, 1993
I.361 *B-ISDN ATM Layer Specification*, 1993
I.362 *B-ISDN Adaptation Layer (AAL) Functional Description*, 1993
I.363 *B-ISDN ATM Adaptation Layer (AAL) Specification*, 1993
I.364 *Support of Broadband Connectionless Data Service on B-ISDN*, 1993
I.371 *Traffic Control and Congestion Control in B-ISDN*, 1993
I.374 *Network Capabilities to Support Multimedia Services*, 1993
I.413 *B-ISDN User–Network Interface*, 1993
I.432 *B-ISDN User–Network Interface — Physical Layer Specification*, 1993
I.555 *Interworking between Frame Relaying Networks and B-ISDN*, 1993
I.580 *General Arrangements for interworking between B-ISDN and 64 kbit/s-based ISDN*, 1993
I.610 *ISDN Operation and Maintenance, Principles and Functions*, 1993
G.804 *ATM Cell Mapping into PDH*, 1993
Q.2010 *General Introduction to Signalling in B-ISDN*, 1993
Q.2020 *Metasignalling Protocol*, 1993
Q.2100 *Signalling ATM Adaptation Layer (SAAL) Overview Description*, 1993
Q.2931 *B-ISDN Digital Subscriber Signalling (DSS2) User–Network Interface Layer 3 Specification for Basic Call/Connection Control*, 1993
[3] ITU-T (CCITT) recommendations
I.121 *Broadband Aspects of ISDN*, 1991
I.321 *B-ISDN Protocol Reference Model and its Application*, 1991
[4] Ahmadi H and Denzel W E, *A survey of modern high-performance switching techniques*, IEEE J. on Selected Areas in Communications **7**, 1091–1103, 1989
[5] Newman P, *ATM technology for corporate networks*, IEEE Communications Magazine, 90–101, April 1992
[6] ATM Forum, *ATM Data Exchange Interface (DXI) Specification, Version 1.0*, August 1993
ATM Forum, *B-ISDN Inter Carrier Interface (B-ICI) Specification, Version 1.0*, August 1993
[7] ATM Forum, *ATM User–Network Interface Specification, Version 3.1*, July 1994

1.2   THE PHYSICAL LAYER

An important parameter of an ATM system is the size of a cell. There are a few of qualitative arguments supporting smaller or larger cell sizes.

- The relative overhead of the cell header is smaller for longer cells.
- The frequency at which headers have to be processed is lower for longer cells.
- As a consequence of elementary queueing theory cell buffers scale with cell size.
- Packetisation delay and switching delay scale with cell size.

The last point probably had the largest impact on the final decision on cell size. According to CCITT recommendation G.164 the total delay for speech traffic must be kept below 25 ms, if no echo cancellation is being used. The actual cell size with a payload of 48 octets results in a packetisation delay of 6 ms. According to CCITT recommendation Q.551, the average delay in a switch should be kept below 450 us. This value will probably not be exceeded by ATM switches based on the current cell format: at a rate of 155 Mbit/s the buffering of 100 cells induces a delay of 270 µs.

It has already been stated that any type of transmission system can be used to carry cells. However, in order to do so, a description is necessary of how cells are inserted and retrieved from a transmission system. Those functions have been collected in a sublayer of the physical layer referred to as the transmission convergence sublayer. The underlying physical medium dependent (PMD) sublayer specifies the characteristics of the physical medium, in particular the type of physical interface at the S_B- or T_B-reference points. Issues addressed are:

- electrical/optical interface
- interface code
- provision of bit timing.

These aspects are mostly covered by older recommendations (such as CCITT G.703) and do not give much insight into ATM technology. As a counterexample one might consider recent descriptions of the physical layer for twisted pair based ATM connections issued by the ATM Forum. In the sequel a short presentation of the physical medium dependent sublayer of a 51.84 Mbit/s interface is given. This interface has already been used for the implementation of a PON access system [3].

Access to B-ISDN via Passive Optical Networks. Edited by U. Killat
© 1996 John Wiley & Sons Ltd

1.2.1   Physical medium dependent sublayer at 51.84 Mbit/s

In [2] a PMD sublayer specification has been developed for using Category 3 Unshielded Twisted Pair (UTP) cabling. Using a pair of wires for each direction a distance of 100 m can be bridged. The transmission scheme used is that of a bandwidth-efficient 16-carrierless amplitude modulation/phase modulation (CAP) technique similar to the well known quadrature amplitude modulation (QAM) technique. The transmitter structure is shown in Figure 1.2.1.

The data first pass a self-synchronising scrambler, the generating polynomial of which is of degree 23. To ensure that the signal in one direction is uncorrelated with that of the other, two different generating polynomials are used, one for each direction of transmission.

In the encoder four data bits are mapped into one of the 16 symbols of the signal constellation of the 16-CAP scheme. The first two bits identify the quadrant, the other two the signal within this quadrant. From the signal stream two streams are derived which are sent to pass-band in-phase and quadrature filters, respectively. At the output of the filters the signals are subtracted and the result is passed through a low-pass filter. As a result of this signal processing, the spectrum of the output signal is confined to a band ranging from 3 MHz to 23 MHz (−10 dB limits).

The remaining part of Chapter 1.2 is devoted to the functions of the transmission convergence sublayer, which can be summarised as follows (see Figure 1.1.3):

— Cell rate decoupling
— HEC generation/verification
— Cell delineation
— Transmission frame adaptation.

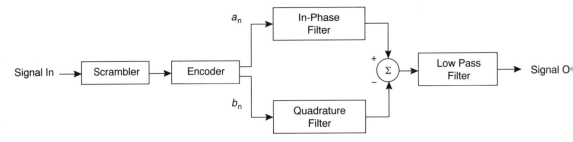

Figure 1.2.1
Structure of a 16-CAP transmitter

1.2.2 Cell rate decoupling

As already mentioned in Chapter 1.1, idle cells will be inserted to adapt the cell stream to the rate of the transmission system. At the receiving side idle cells will be discarded. The ATM layer is not aware of the existence of idle cells. The process of idle cell insertion/discard is known as cell rate decoupling. The header structure of an idle cell is shown in Figure 1.3.10.

1.2.3 HEC generation/verification

The header field of an ATM cell is protected by a CRC using the generator polynomial $x^8 + x^2 + x + 1$. This procedure is called header error control (HEC).

The HEC is capable of

- correcting single bit errors
- detecting a couple of multiple-bit errors.

The information contained in the header field is used for the routing of cells. A misrouting will lead to

- a missing cell in the considered connection
- a cell coming out of the blue in another connection.

Upon the detection of an erroneous cell, this cell will be discarded. Thus the second unwanted effect is avoided. According to the algorithm used, the decoder is switched between an 1-error correction mode and a multiple-error detection mode according to the state diagram depicted in Figure 1.2.2.

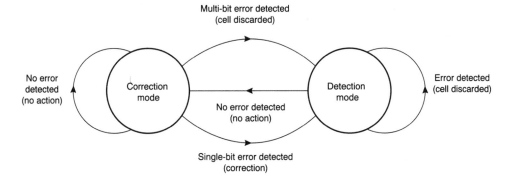

Figure 1.2.2
State diagram of the decoder of the HEC code

This strategy of the receiver matches the error statistics of fibre-based transmission systems, where single-bit errors are predominant but where long bursts of errors may also be observed at a far lower probability.

1.2.4 Cell delineation

One of the problems of a receiver in an ATM-based system is the detection of cell boundaries. Unlike the situation in classical synchronous data communication systems no flags are foreseen to identify the start and end of a cell. An additional requirement was to base the cell delineation process only on the cell structure and not on framing information that may or may not exist in a given transmission system.

One reasonable solution would have been to base the delineation process on the detection of idle cells. In particular when spacers are introduced at the outlets of network nodes (see Section 3), synchronisation information would occur sufficiently frequently. The solution chosen by the ITU-T is based on the HEC mechanism. In the absence of bit errors consecutively correct HEC values are indicative of a correct estimate of the cell boundaries. So hunting for a zero

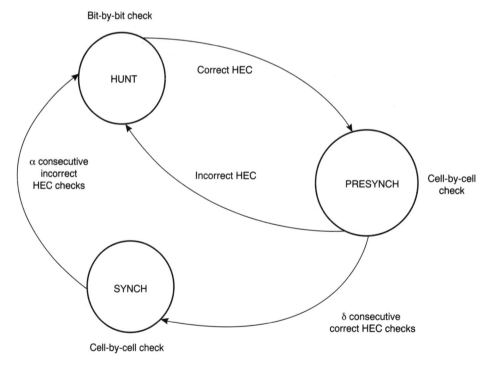

Figure 1.2.3
State diagram of the cell delineation process

syndrome after the reception of five bytes leads the receiver to the correct synchronisation. The corresponding state diagram is shown in Figure 1.2.3.

The receiver starts in the HUNT state and performs a check on the assumed header field boundaries. These are shifted bit-wise until a correct HEC leads to the PRESYNCH state. If the correct HEC can be confirmed δ times the SYNCH state is reached. It will be left only after α incorrect HEC values. Currently the values $\delta = 6$ and $\alpha = 7$ are in use.

For the cell based physical layer (see Chapter 1.2.6) a scrambler called the Distributed Sample Scrambler is used which covers the full cell (header and payload). In order still to be able to base the cell delineation on the HEC mechanism, in this case the HEC field has to be corrected to account for the scrambler bits added to the first four octets of the cell. For all the other transmission systems the self-synchronising scrambler based on the polynomial $x^{43} + 1$ is used to avoid malfunctioning of the cell delineation process owing to a critical data pattern in the payload of the cell. This scrambler only covers the payload field. As this scrambler has only two taps error multiplication is minimised.

1.2.5 Cell transport via an STM-1 system

Among the different possibilities for carrying ATM cells, the use of Synchronous Digital Hierarchy (SDH) transmission systems was the first to become standardised (recommendation I.432 [1]). Cells are carried in a so-called VC4-container of an STM-1 transport system (Figure 1.2.4).

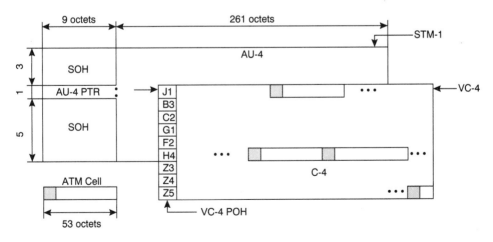

Figure 1.2.4
Transport of ATM cells in an STM-1 transport system

Cells are byte-aligned to the byte-structure of the container. As its payload is not an integer multiple of the cell size, a cell can be spread over more than one container–as indicated in the right bottom corner of the figure. The so-called AU4-pointer points to the start of the VC4-container.

Besides the STM-1 system (155.52 Mbit/s), in North America the so-called STS1 (51.84 M/bits) system of the SONET family of transmission systems has become very popular. The support of the STS1 system is being described at the physical medium sublayer by the ATM Forum specifications [2].

1.2.6   Cell based transmission at 155.520 Mbit/s

Both the CCITT and the ATM forum present an option of carrying an infinite stream of cells over a transmission medium without any additional framing structure. The transmission system merely provides a bitstream over which ATM cells are transmitted without any space between them. The ATM forum [2] does specify transmission over fibre and twisted pair. In both cases an 8B/10B block code is applied to guarantee a negligible low frequency content and a sufficient amount of transitions in the signal.

REFERENCES

[1] ITU-T (CCITT) recommendation
[2] ATM Forum, *ATM User–Network Interface Specification, Version 3.1*, July 1994
[3] Hermann D D, Huang G, Im G-H, Nguyen M-H, Werner J-G and Wong M K, *Local Distribution for Interactive Multimedia TV to the Home*

1.3 THE ATM LAYER

The ATM Layer is concerned with the transport of cells between users of the ATM layer. The format of a cell at the user network interface (UNI) is depicted in Figure 1.3.1. As the ATM layer follows the paradigm of connection orientation, connections have to be set up and quality of service parameters have to be negotiated for these connections. The cell routing information contained in the header of a cell is a label, not an explicit address. The label uniquely identifies a connection on a given link. All routing information is for connections and therefore linked to the label values.

The functions of the ATM layer include

- multiplexing and switching of ATM connections
- identification of certain cell parameters
- traffic shaping

In this chapter mainly the first aspect will be discussed; the last one will be taken up in Chapter 1.6.2.

1.3.1 ATM connections

ATM connections are either permanent virtual circuits (PVCs) or switched virtual circuits (SVCs). In the former case they are installed by network management actions, in the latter as the outcome of a signalling protocol. The signalling protocol defined by the ITU-T is described in the recommendation Q.2931 [1]. In any case the establishment of a connection can be seen as a sort of contract.

The network provider will guarantee certain quality of service (QoS) parameters such as cell delay, cell delay jitter, or cell loss rate. The network user will characterise the data stream to be carried via the connection by a set of traffic parameters such as peak rate, sustainable cell rate, and burst tolerance (see also Chapter 1.5). Usage parameter control (discussed in Chapter 1.6) is a function which measures whether or not the customer is compliant with his contract.

Connections can be point-to-point, point-to-multipoint, and (under discussion) multipoint-to-multipoint. ATM connections guarantee the in-sequence delivery of cells. Being distinguished by the aforementioned label, different ATM connections can be multiplexed on one physical link. For the label an address space of 24 bits has been reserved (at UNI) which is subdivided into a virtual path identifier (VPI) of 8 bits and a virtual channel identifier (VCI) of 16 bits (see Figure 1.3.1). At the network node interface (NNI) a slightly different cell format is used: the GFC field is missing and the VPI field is in turn enlarged to a length of 12 bits.

As a consequence of the distinction between VPI and VCI a hierarchy has been established: one virtual path (VP) can contain up to 2^{16} virtual

Access to B-ISDN via Passive Optical Networks. Edited by U. Killat
© 1996 John Wiley & Sons Ltd

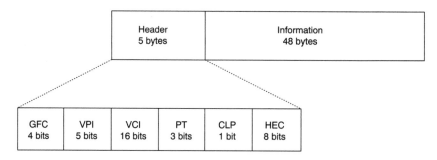

GFC : Generic Flow Control   CLP : Cell Loss Priority

VPI : Virtual Path Identifier   HEC : Header Error Control

VCI : Virtual Channel Identifier

PT   : Payload Type

Header 5 bytes	Information 48 bytes

GFC 4 bits	VPI 5 bits	VCI 16 bits	PT 3 bits	CLP 1 bit	HEC 8 bits

Figure 1.3.1
Format of a cell

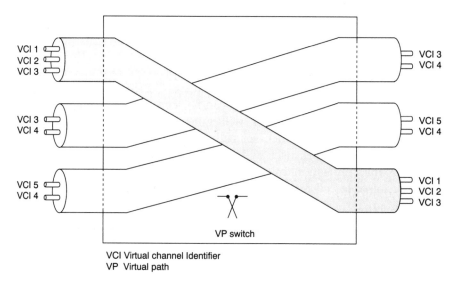

VCI Virtual channel Identifier
VP Virtual path

Figure 1.3.2
VP switching

channels. The impact of this label structure becomes obvious when switching is considered: There are VC switches and VP switches. As is illustrated in Figure 1.3.2, a VP switch will leave the content of the VCI field unchanged, whereas a VC switch may change both the VPI and the VCI.

VP connections therefore give the appearance of leased lines which carry a multiplex of connections. Virtual paths thus enable the network operator economically to offer virtual networks with features of closed user groups. Moreover, the network provider will benefit from the possibility of creating a two-level hierarchy in his network in a natural way. From a management point of view it is important to stress that a certain quality of service is associated with a given VP and this quality of service will be inherited by the VCs contained in the VP.

1.3.2 ATM switching

The generic function of a switch is to take data from an input port and to release them at an output port. For an ATM switch, in addition to this function, a label switching is also necessary, which is typically performed in a so-called translator (see Figure 1.3.3). In an abstract manner the switching of a cell can be described as a mapping

$$(\text{port}_{in}, (\text{VPI/VCI})_{in}) \rightarrow (\text{port}_{out}, (\text{VPI/VCI})_{out}).$$

The translator contains a table with pairs $(\text{VPI/VCI})_{in}$ as an address to retrieve the value of the pair $(\text{VPI/VCI})_{out}$. The need to translate VPI/VCI values stems from the fact that a VPI/VCI value is unique only on a given link. Therefore, it may happen that connections with identical VPI/VCI values but different input links have to be switched to the same output link. In order to be able to distinguish these connections on the output link, a label translation is necessary.

Another lesson can be learned from Figure 1.3.3. Suppose that the two cells with labels b and c have to be switched to the same output port. During the time unit of one cell only one of the two cells can be forwarded to the destination output port; the other one has to wait in a buffer. Thus ATM switches are different from conventional crossbar systems in that they need (a substantial amount of) buffering. The waiting time of a cell in buffers of the switch is a random variable. Therefore time transparency is lost which leads to the "asynchronous" behaviour already indicated in the term "asynchronous transfer mode".

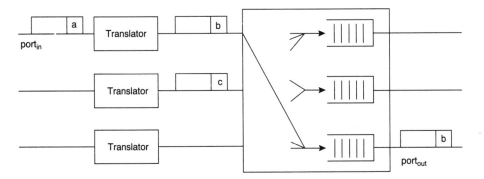

Figure 1.3.3
Generic ATM switch; a, b, c: VPI/VCI-values

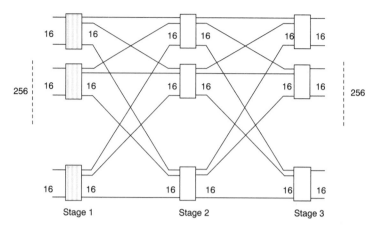

Figure 1.3.4
Link system

In Figure 1.3.3 output buffers have been provided for the switch fabric shown. This is only one of a couple of alternatives in constructing ATM switches; in the early phase of ATM development many investigations concentrated on principles and performance of switch architectures; an overview on the topic is given in. [2] Switching networks in general exhibit a hierarchical design. There do exist switching matrices of a limited size, e.g. 8×8. Typically a switching matrix preserves cell sequence integrity. Greater networks are constructed from elementary matrices according to the principle of link systems. A link system consists of a couple of "stages" each of which represents a column of switch matrices. Between switching stages a wiring scheme is used which includes at least as many links as there are inputs to the link system (Figure 1.3.4).

Increasing the number of internal links ("expansion"), and thereby the number of possible paths from a source to a destination, has almost the same effect as working with no expansion but at a correspondingly increased speed of the internal links.

1.3.2.1 Matrix switches

For the elementary matrix switches of dimension $m \times m$, cross-bar systems with input or output buffering have frequently been proposed (see Figure 1.3.5a,b).

The advantage of input buffering is that the buffer is written and read at the line rate. The disadvantage of input buffering is the head of line blocking: if the first cell in the input queue is blocked because its corresponding output is busy, any other cell after the head of line cell will be blocked as well — even if its corresponding output is free. For this reason, a pure input buffering is not implemented.

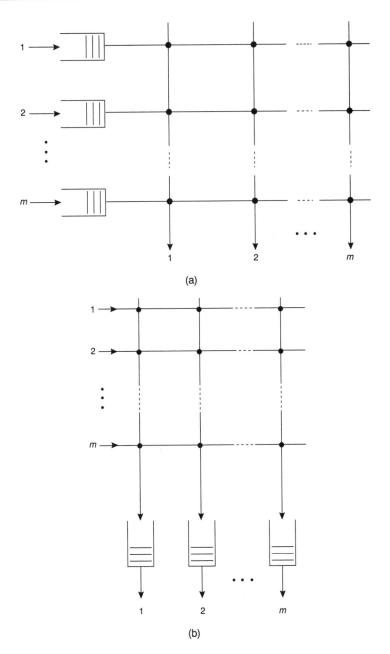

Figure 1.3.5
(a) Input buffering; (b) Output buffering; (c) Central buffering

Output buffering avoids the described effect but suffers from an m times increased writing speed. Output buffered switches are considered as generic switch models, as they demonstrate an ideal first in first out (FIFO) behaviour. A variant of the output buffered switch is the central

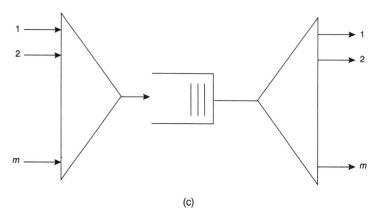

(c)

Figure 1.3.5 (*continued*)

queueing approach where logical output queues are implemented in a central memory (Figure 1.3.5c).

For such an approach write and read cycles have to be executed at an m times higher speed compared with the speed of the input lines. The waiting time behaviour of the central queueing system is identical to that of the output queueing system. However, as several queues are realised within a single physical memory, advantage can be taken from an averaging effect. Actually, a memory reduction of a factor of roughly $m^{1/2}$ is obtainable. But it must not be forgotten that a complicated buffer management is necessary in order to guarantee that each cell safely arrives at the desired output port.

For the analysis of a switching network one of the most frequently used assumptions is that packet arrivals on the m inputs obey independent and identical Bernoulli processes. If a matrix switch with output queueing is investigated on this basis it turns out that the behaviour of an arbitrarily chosen output queue is similar so that of a classical M/D/1 system. [7] In fact the average queue lengths L of both systems differ only by a factor $(m-1)/m$:

$$L_{\text{Output}} = \frac{m-1}{m} L_{\text{M/D/1}}$$

A similar expression holds for the average waiting times W:

$$W_{\text{Output}} = \frac{m-1}{m} W_{\text{M/D/1}}$$

The similarity of both systems in terms of mean waiting time is shown in Figure 1.3.6 (see [9]) where the M/D/1 limit is obtained for $m \to \infty$.

If a switching matrix is to be realised as a single chip, then for today's silicon technology pin count, chip area and power dissipation will be constraints which ensure that at line speeds of about 155 Mbit/s m will not be substantially greater than 16. The speed increase of m for

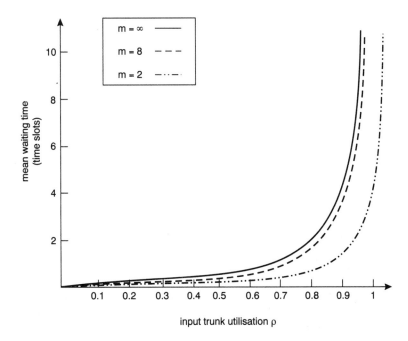

Figure 1.3.6
Output queueing: Mean waiting time as a function of trunk utilisation

writing to an output buffer is in practical implementations reduced to values smaller than m. This is justified by the assumption that simultaneous arrivals on all input lines for one output line are highly improbable. [3,4]

1.3.2.2 Link systems

Taking the number of cross-points NCP as a measure of complexity, for a matrix switch with m inputs NCP is given by

$$NCP = m^2$$

If a switch with $N \gg m$ inlets has to be constructed, link systems are used, as already stated above. The minimal NCP that can then be obtained for a link system with only one route from one input to one output is given by

$$NCP = N \cdot m \cdot \log_m N < N^2$$

Thus the main advantage of link systems compared with an extended matrix structure is their reduced hardware effort. In addition, many link systems have more than one route from one input to one output. This redundancy can be exploited to achieve a graceful degradation in case of failures.

As already indicated above, the dimensioning of buffers relies to a great extent on queueing theory. Unfortunately, the standard assumptions of uniform and geometrically distributed traffic at the input ports do not hold in realistic traffic scenarios. To circumvent this problem, an advanced switch design was proposed [5,6] which consists of three different functional blocks: a randomising block, a routing block and a resequencing block (Figure 1.3.7). The first two blocks consist of one or more switching stages.

The randomising block will deliver a uniform traffic distribution to the routing block. An example for this input traffic shaping is shown in Figure 1.3.8. [8] Here a link system such as the one shown in Figure 1.3.4 is considered. A wedge-shaped input pattern with utilisation $\rho = 1$ on input 1 and $\rho = 0$ on input 256 has been investigated. The output port is chosen randomly for every cell. The results in Figure 1.3.8a refer to a network consisting of stages 2 and 3 only. The losses follow the wedge shape of the input traffic. In Figure 1.3.8b the system is equipped with an additional first input stage working in a randomising mode. Obviously, the loss behaviour could substantially be improved and the traffic profile at the inputs is no longer visible in the rest of the network. In Figure 1.3.7 the routing block and its buffers can be dimensioned using standard queueing theory methods. Owing to the randomly chosen paths and associated random delays, the in-sequence delivery of cells is no longer guaranteed. For this reason a resequencing buffer is foreseen which reconstitutes the correct order. To this end a straightforward time-stamp resequencing algorithm is employed. The different delays experienced by different cells on their way through the randomising

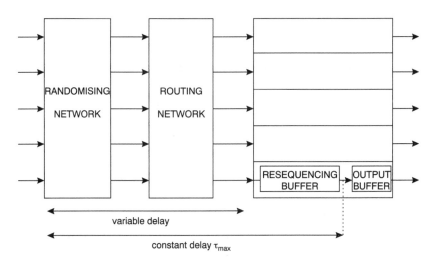

Figure 1.3.7
Randomising multipath network with resequencing

and routing blocks are adjusted in the resequencing buffer to a maximum delay τ_{max}. If t_0 was the time stamp at the entrance of the switch then $t_0 + \tau_{max}$ denotes the point in time when the cell will have to leave the resequencing buffer. The resulting buffer needed in the resequencing block equals the sum of all buffers along any path across the randomising and routing blocks. It should be mentioned that the number of buffers in the randomising stages can be fairly small and — depending on the implementation of the randomising function — can even be reduced to zero. [8]

1.3.2.3   *Routing principles of a switch*

From $((VPI/VCI)_{in}, port_{in})$ the $port_{out}$ can be determined. There are two major principles that determine how this routing decision is implemented in the switch:

- table-routing
- self-routing

If table-routing is applied (see Figure 1.3.9a) then every switch element (every elementary matrix switch) of a switching network maintains a table per input port stating to which of *its* output ports a particular VPI/VCI shall be routed. The concatenation of the routing decisions in the individual elementary matrix switches represents the route of a given VPI/VCI across the total switch. The benefits of this table-routing are:

- the cell format contains no additions and remains compliant with the standard
- for small systems where VPI/VCI values are unique in the whole system no other element is necessary to construct a switch.

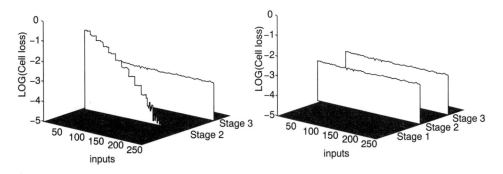

Figure 1.3.8
Effect of a wedge shaped traffic input profile on cell loss. (a) 2 stage network; (b) 3 stage network with randomising input stage

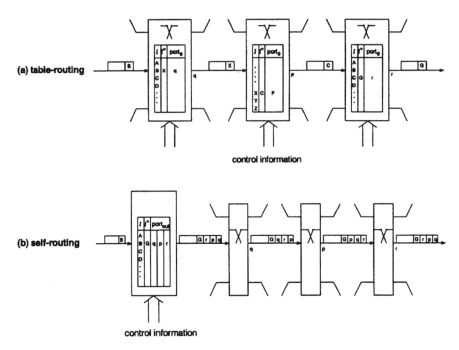

Figure 1.3.9
Allocation of routing decisions in the switching network. (a) table-routing; (b) self-routing

The drawback of table routing is that at call set-up or at call release the individual routing tables of all elementary switches have to be updated by a switch control module via some control channel. In contrast, if self-routing is applied one exploits the fact that all routing information is already available in the translators before the switching network (see Figure 1.3.3). Therefore the translator, besides producing the (VPI/VCI)$_{out}$ pair, also generates the port$_{out}$ and encodes it as the sequence of routing decisions in the k stages of the switch:

$$(\text{port } x_1, \text{port } x_2, \ldots, \text{port } x_k)$$

This information is placed in a "tag field" in front of a cell (Figure 1.3.9b). In the self-routing switch that follows, in every stage the relevant subfield of the "tag field" is read and the cell is routed accordingly.

To summarise: in small switching systems table routing is a valuable option; for extended systems self-routing is the preferred approach.

1.3.3 Identification of special cell types

As stated in Chapter 1.1, only valid cells are passed to the ATM layer. The cell header (see Figure 1.3.1) already contains information that indicates that some cells need special treatment in the ATM layer. One

	Byte 1	Byte 2	Byte 3	Byte 4
Reserved for Physical Layer	PPPP 0000	0000 0000	0000 0000	0000 PPP1
OAM cell at Physical Layer	0000 0000	0000 0000	0000 0000	0000 1001
idle cell	0000 0000	0000 0000	0000 0000	0000 0001
unassigned cell	AAAA 0000	0000 0000	0000 0000	0000 AAA0

	VPI		VCI	
Metasignalling Channel	0000 0000	0000 0000	0000 0001	
Broadcast Signalling Channel	0000 0000	0000 0000	0000 0010	

Figure 1.3.10
Preassigned cell header values

obvious example is the unassigned cells (see Figure 1.3.10). These cells do not carry information in the payload. But in contrast to the idle cells in the physical layer, they have a VPI/VCI value. Most cells with special header values are related to management actions. For example, the ATM Forum has assigned special VPI/VCI values to functions like meta-signalling, broadcast signalling, operation and maintenance (OAM) flows. The F4 flows, for example, exhibit a special VCI value and a VPI value identical to that of the virtual path to be controlled. F5 flows are used to control virtual channels. Therefore they have a VPI/VCI field identical to that of the virtual channel to be controlled; the indicator for the F5 flow is then found in the payload type (PT) field. The actual semantics of the OAM message (e.g. alarm indication, far-end receive failure, loopback procedures) is coded in the payload of an OAM cell.

In general, the payload type field allows one to distinguish between user and non-user cells. Two functions of the payload type field are related to user cells.

- The PT field allows one to differentiate between two types of service data unit (SDU) — a feature which is exploited in the construction of the AAL-5 protocol described in Chapter 1.4.

- The occurrence of a congestion situation can be indicated in the PT field.

User cells are distinguished according to a priority level (cell loss priority bit). Cells of low priority are discarded in congestion situations. The usage parameter control (UPC, see Chapter 1.6) may use this mechanism to "tag" cells which do not conform to the agreed connection contract.

It should be noted at this stage that the cell loss priority mechanism applies to user cells within one connection. Its primary area of application is related to connections carrying video signals, where, for example, synchronisation information needs a top priority. If different qualities of service for different connections need to be expressed, this has to be done at the VP level, and the network provider will make use of dedicated resources (e.g. buffers) to guarantee a certain grade of service to certain VPs.

The cell header at the UNI is different from that at the network network interface (NNI) in that it exhibits the generic flow control (GFC) field. The GFC field has only local significance. The GFC mechanisms are not yet fully described. They are used to alleviate short-term overload conditions that may occur when customer premises networks are connected at the UNI. GFC information can be transported by assigned or unassigned cells. The GFC protocol must assure that all terminals contending for the UNI get access to that resource according to their connection contracts. Thus only traffic to the ATM network and not that from the ATM network is controlled. In general, no flow control is foreseen at the ATM layer.

REFERENCES

[1] ITU-T (CCITT) recommendation
    Q.2931 *B-ISDN Digital Subscriber Signalling (DSS2) User–Network Interface Layer 3 Specification for Basic Call/Connection Control*, 1993
[2] Ahmadi H, Denzel W E: A Survey of Modern High-Performance Switching Techniques, IEEE J. on Selected Areas in Communications 7, pp. 1091–1103 (1989)
[3] Yeh Y S, Hluchyi M and Acampora A, *The knockout switch: a simple, modular architecture for high-performance packet switching*, IEEE J Selected Areas in Communications, 5, 1274–1283, 1987
[4] Killat U, Kowalk W, Noll J, Keller H G, Reumermann H J and Ziegler U, *A Versatile ATM Switch Concept*, Proc. XIII International Switching Symposium, Stockholm, paper A 6.4, June 1990
[5] Henrion M A, Schrodi K J, Boettle D, de Somer M and Dieudonné M, *Switching Network Architecture for ATM Based Broadband Communications*, Proc. XIII International Switching Symposium, Stockholm, paper A 7.1, June 1990
[6] Petit G H, Buchheister A, Guerrero A and Parmentier P, *Performance Evaluation Methods Applicable to an ATM Multi-Path Selfrouting Switching Network*, Proc. 13th International Teletraffic Congress, 917–922, 1991
[7] Gross D and Harris C M, *Fundamentals of Queueing Theory*, John Wiley & Sons: New York
[8] Schmidt K and Killat U, *Performance of a traffic smoothing switching element*, (to be published) in: Eur. Trans. Telecom.
[9] Karol M J, Hluchyj M G and Morgan S P, Input versus output grieveing on a space-division packet switch, *IEEE Jr. Comm*, 35, 1347–56, 1987.

1.4 ATM ADAPTATION LAYER (AAL)

The ATM layer offers a service to its users which is characterised by

- connection orientation
- atomic data units of the format of a cell
- a certain quality of service which includes some delay jitter.

This service is, seen from an applications perspective, also a set of restrictions. To bridge the gap between what is offered and what is asked for, a couple of adaptation layers have been created [i]. The starting point was four service classes which were felt to represent a sensible amount of future B-ISDN applications. These are:

Class A: Circuit emulation
Class B: Variable bit rate service with time synchronisation between sender and receiver
Class C: Connection oriented data service
Class D: Connectionless data service

The main distinction between the connectionless and the connection oriented data services is that the former does not require an associated signalling procedure. However, this is a feature which mainly concerns the control plane and not the AAL layer *per se*. As a result of this argument, protocols having been specified for Classes C and D, namely AAL-3 and AAL-4 protocols, were merged into one type AAL-3/4. Similarly, the AAL Type 5 protocol developed as a simplified version of AAL-3 is not bound to connection oriented data services only.

As a general rule, the AAL layer is subdivided into two sublayers

- the segmentation and reassembly (SAR) sublayer
- the convergence sublayer (CS).

The SAR takes the data units provided by the CS and segments them into fixed length cell payloads. Conversely, at the receiving side the original formats are reconstructed from the cell stream.

The CS performs additional functions which are different for different AAL types and have to do, for example, with multiplexing and timing recovery.

1.4.1 ATM adaptation layer Type 1

The most important application of an AAL Type 1 protocol is circuit emulation. This applies to all signals of the PDH hierarchy as well as to those of the SDH hierarchy. The AAL in these cases receives and

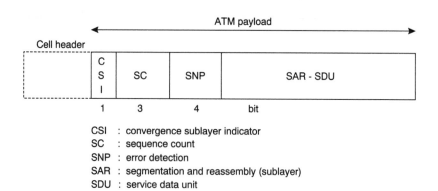

CSI : convergence sublayer indicator
SC : sequence count
SNP : error detection
SAR : segmentation and reassembly (sublayer)
SDU : service data unit

Figure 1.4.1
Structure of an AAL Type 1 frame

delivers bitstreams. The AAL has to hide two effects which result from the ATM principles: cell loss and cell delay variation. For the former a sequence numbering mechanism is offered; for the latter, methods for source clock recovery. The structure of an AAL-1 frame can be seen in Figure 1.4.1.

The segmentation and reassembly protocol data unit (SAR-PDU) header consists of a 4-bit sequence number and a sequence number protection field detecting all 1-bit and 2-bit errors. Thus the CS can identify lost and misinserted cells to a certain degree. To be precise, the sequence number consists of two parts: a sequence count (SC) of 3 bits and a convergence sublayer identifier (CSI) of 1 bit, the latter is being used in a rather sophisticated manner: if the sequence count is even, the CSI bit will indicate the existence of a pointer which would point to some byte in the payload where a "frame" starts. This feature can be used to transport structured data such as, for example, a PCM-30 frame. If the sequence count is odd, the sequence of CSI bits will be used to encode the frequency of the source clock which has to be regenerated at the receiver. This method is referred to as the synchronous residual time stamp (SRTS) method.

The cell delay variation will necessitate a play-out buffer at the receiver. If that buffer is emptied at a rate corresponding to the sender's clock it will be able to compensate for the delay jitter in the network. One implicit method of doing this is to control the rate at which the buffer is emptied in such a way that neither buffer overflow nor buffer starvation occurs. An explicit method of conveying the sender clock to the receiver is the aforementioned SRTS method. This method works as follows.

The time needed for s cycles of the service clock f_s defines a time interval T. During this time interval the number n of cycles of the network clock f_n of the B-ISDN is counted. This count n will differ from an ideal nominal value n^*. The difference value $\Delta n = (n - n^*)$ is transmitted to the receiver via the CSI bit. The receiver upon reception of Δn, and hence n, can derive the clock frequency of the sender by evaluating the following relationship:

$$f_s = \frac{s}{n} \cdot f_n$$

Finally, it should be mentioned that within AAL-1 forward error control (FEC) techniques have also been specified — mainly for applications containing video signals. The FEC methods used rely on Reed — Solomon codes and interleaving techniques.

1.4.2 ATM adaptation layer Type 2

The adaptation layer Type 2 is concerned with variable bit rate data which — as in the case of circuit emulation — need time transparency. Work on this adaptation layer has not progressed far enough to make a detailed description worthwhile.

1.4.3 ATM adaptation layer Type 3/4

This adaptation layer specifies connectionless and connection oriented transfer of data frames via an ATM network. The CS is for AAL-3/4 and AAL-5 further subdivided into the service specific convergence sublayer (SSCS) and the common part convergence sublayer (CPCS), see Figure 1.4.2.

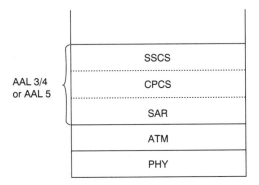

SSCS : Service Specific Convergence Sublayer
CPCS : Common Part Convergence Sublayer
SAR   : Segmentation and Reassembly Sublayer

Figure 1.4.2
Structure of the ATM adaptation layer for Types 3/4 and 5

The SSCS for connectionless services typically is a null-functionality whereas for connection oriented application such as, for example, the B-ISDN signalling, the SSCS sublayer will contain functions for in-sequence delivery, error handling, flow control amongst others (see Chapter 1.4.5). At this stage only CPC and SAR functionalities are further considered. The functions of AAL-3/4 can briefly be summarised as follows. Variable length data packets (in the range of 1 to 65 535 bytes) of the application are padded to multiples of 32 bits and are equipped with a header and a trailer. The resulting CS-PDU is then subdivided into segments of 44 bytes, each of which receives its own header and trailer to form an SAR-PDU which is then transported in the payload of an ATM cell.

1.4.3.1   The common part convergence sublayer (CPCS)

This sublayer contains basic functionalities needed to build up connectionless protocols (e.g. connectionless network access protocol (CLNAP)) or connection oriented protocols (e.g. Frame Relay service on top of ATM). These functions are:

- identification of the transported CPCS-SDU
- error handling
- buffer allocation.

These functions can better be understood when the structure of the CPCS-PDU is considered (see Figure 1.4.3).

The CPCS-PDU consists of a 4-byte header, an information field of variable length and a 4-byte trailer. The header and trailer contain fields

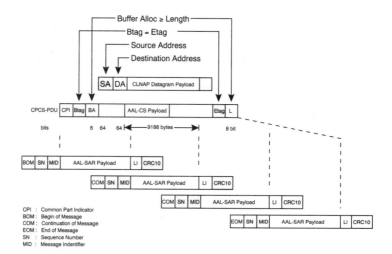

Figure 1.4.3
Segmentation of messages according to AAL Type 3/4 protocol

which do correspond to each other: B(egin)TAG and E(nd)TAG, buffer allocation (BA) and length (L). The values of BTAG and ETAG have to be identical and identify a given CPCS-PDU. A subsequent CPCS-PDU has to have a different BTAG/ETAG value. Differing values of BTAG and ETAG for a received CPCS-PDU will result in an error message being sent to the application. The BA field indicates to the receiving station the amount of buffer necessary to receive properly the AAL-SDU. As one SDU can also be transported in more than one CPCS-PDU the condition

$$BA \geq L$$

holds, where L describes the actual length of the CPCS-PDU. A violation of this inequality results in an error situation. The values of BA and L are given in bytes unless something else is specified in the CPI field.

1.4.3.2   *The segmentation and reassembly sublayer*

According to Figure 1.4.3 the CPCS-PDU (= SAR-SDU) is segmented into units of 44 bytes and put into the payload of an SAR-PDU. The header/trailer information contained in an SAR-PDU supports the following functions of the SAR sublayer:

- identification of the SAR-SDU
- error handling
- sequence integrity
- multiplexing of CPCS connections.

Identification of an SAR-SDU is supported by the segment type (ST) field which determines whether a given SAR-PDU represents the beginning (BOM), a continuation (COM) or the end (EOM) of a message. A missing or misinserted SAR-PDU will be detected by evaluation of the 4-bit sequence number (SN) field. No resequencing is needed because the ATM service preserves the sequence. All SAR-PDUs belonging to the same CPCS-PDU are characterised by the same multiplexing identifier (MID). Thus, simultaneous and interleaved transmission of messages is possible as each is distinguished by its MID value. It is important to note that each SAR-PDU is protected by its own CRC10. This is a protection of the header information but also of the payload. This point is crucial when connectionless services are considered. As indicated in Figure 1.4.3, the addresses of the connectionless service protocol are contained in the corresponding BOM cell and are protected by the CRC10, which makes this address information reliable.

1.4.4 ATM adaptation layer Type 5

1.4.4.1 *The common part convergence sublayer*

As already stated, AAL Type 5 can be considered as a simplified version of Type 3/4. Padding of CPCS-SDUs is done in multiples of 48 bytes. A CPCS-PDU is segmented into packets of 48 bytes which fit exactly into the payload of a cell. All control information of the CPCS sublayer is found in the CPCS trailer. The functions of the CPCS are:

- identification of the transported CPCS-SDU
- user-to-user information transfer
- error handling

These functions will be explained with reference to Figure 1.4.4 which displays the format of a CPCS-PDU. The CPCS-SDU is contained in the CPCS-PDU payload. The length field and the CRC-32 assure that only correctly received CPCS-PDUs are further processed. The CPCS-UU (UU: user-to-user) field enables the users to directly exchange information — a possibility which is not provided in AAL-3/4.

1.4.4.2 *The segmentation and reassembly sublayer*

The main function of this layer is to detect the beginning and end of an SAR-SDU (= CPCS-PDU). In AAL-3/4, this functionality was realised by using the segment type field of the SAR header. In AAL-5 such a header is missing. To execute the desired function one bit of the payload type field of the cell header is (mis-)used. This bit indicates whether the following SAR-PDU contains the end of an SAR-SDU. An explicit identification of the beginning of an SAR-SDU is not provided.

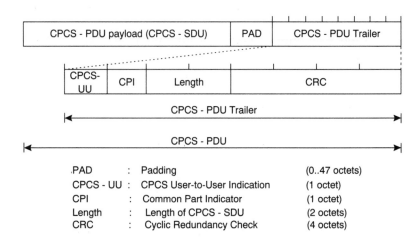

Figure 1.4.4
Format of a CPCS-PDU of the AAL Type 5 protocol

1.4.5   The adaptation layer for signalling (SAAL)

As in the narrowband ISDN, signalling is carried "outband" in dedicated signalling channels. The signalling messages assume a reliable channel between its peer entities. To create this channel a service specific convergence sublayer (the service specific connection oriented protocol (SSCOP)) has been defined which can be put on top of AAL-3/4 or AAL-5 and which will ensure a service that can be considered as the equivalent of the service provided by the (link access) protocol LAP D in the narrowband ISDN. Thus sequence integrity, error correction, flow control and connection management are dealt with.

REFERENCES

[1]   ITU-T (CCITT) recomendations *I.362 B-ISDN Adaptation Layer (AAL) Functional Description*, 1993
I.363 B-ISDN ATM Adaption Layer (AAL) Specification, 1993

1.5   CONNECTION ADMISSION CONTROL IN ATM NETWORKS

As already stated several times in this section, an ATM network works according to the paradigm of connection orientation. This implies that before any communication between the interested parties can start, a connection has to be established. To this end the initiator of the connection talks to the network control which has to check whether network resources are available to support the new call. If this happens to be the case, the call will be accepted; otherwise the network control is said to block the call. The communication with the network control is formalised in a signalling protocol.

In the B-ISDN, signalling is more complicated than in the ISDN. This has to do with the greater flexibility offered in ATM networks which is reflected in the traffic parameters, different AAL types and connection types (e.g. multicast). Therefore, the signalling protocol Q.2931 specified by the ITU-T for B-ISDN is more than a straightforward extrapolation of the ISDN signalling protocol Q.931. In this chapter no description of the Q.2931 [8] protocol will be given. Instead, some problems of call control will be highlighted which are specific to ATM networks and have no counterpart in conventional networks.

To understand the problems involved, a quick review will be given of what resource allocation means for an ISDN network. A set-up message received from a subscriber "A" who wants to build up a connection will inform the local exchange about the number (i.e. the address) of the desired communication partner "B". Usually, one 64 kbit/s channel is necessary to establish the connection. In the network, routing tables are kept which describe a preferred (and an alternative) way from "A" to "B". What the call admission control has to check is whether along the preferred (alternative) route there is always a free time slot on every link. If this condition is met, the call will be accepted — provided subscriber "B" is not busy. How will this situation change in an ATM network? There are a few aspects to consider.

- ATM sources in general do not transmit at a fixed data rate.
- ATM packets change their interarrival times on their way across the network owing to cell delay variation.
- There is a certain probability for cell loss and different services may require a different maximal cell loss probability.
- There is a certain probability that a certain delay (delay variance) is exceeded and different services may require different maximal values.

Access to B-ISDN via Passive Optical Networks. Edited by U. Killat
© 1996 John Wiley & Sons Ltd

Owing to the statistical nature of the sources' traffic parameters and owing to the change of these parameters on the way through the network, resource allocation can no longer be based on counting the amount of available resources but must be based on estimating them. The connection admission control (CAC) must decide, in real time, based on the traffic parameters provided by an incoming connection, whether the new connection can be accepted. Connection acceptance implies that the network provider can guarantee the quality of service requested by that new connection while sustaining the quality of service for the on-going connections. The traffic parameters, referred to as traffic descriptor by the ITU, are part of the traffic contract that both parties (call initiator and network provider) are going to engage in. Concerning the traffic descriptor, the ITU recommendation I.371 only refers so far to the peak cell rate, whereas the ATM Forum [9] in addition to that has defined

- the sustainable cell rate (SCR) and
- the burst tolerance.

The sustainable cell rate provides an upper bound on the conforming average rate of a connection and the burst tolerance limits the time that a source is allowed to send at its peak cell rate. These two parameters are defined in the context of the generic cell rate algorithm (GCRA, see Chapter 1.6).

To simplify the procedures, the network provider may install service classes (most probably linked to VPs) for which a certain quality of service has to be maintained. The discussion of call admission control can then be limited to one parameter set of quality of service, or to simplify the discussion even further, to one parameter (for example, the cell loss probability). The problem of the maintaining an upper bound to cell loss can then be formulated as follows.

If the probability density function of the instantaneous aggregated data rate "X" of all accepted connections and the newly arriving connection is known, then for a bufferless system the overflow probability (that "X" exceeds the capacity "C" of a given VP) can be calculated and identified with the cell loss probability (see Figure 1.5.1).

Unfortunately, it is almost impossible to deduce from the traffic descriptor the resulting probability density function and hence it is also almost impossible to determine its tail. Theorems that guarantee a certain limiting distribution are often not applicable because in most cases the number of (broadband) sources cannot be considered getting close to infinity.

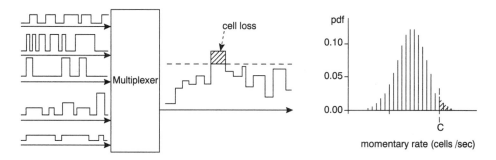

Figure 1.5.1
Statistical multiplexing and cell loss

So far, several algorithms have been proposed for application in connection admission control (see for example [1–7]). One possible approach is peak rate allocation. However, its use doesn't take advantage of a very important characteristic of ATM networks: statistical multiplexing. In fact, with peak rate allocation, the statistical multiplexing gain obtained is null. Other proposed algorithms include the Gaussian approximation, the convolution method, and fluid flow models. Recent studies have also considered the use of neural networks (see [5,6]). Only the first two algorithms will be described below.

1.5.1   Gaussian approximation

With the Gaussian approximation method, connections are described by two parameters: the mean bit rate and the standard deviation. When a new connection to the system arrives, the method has to determine (by using an approximate formula) the bandwidth necessary for all the established connections plus the new one and ascertain whether that bandwidth value exceeds the available link bandwidth. To this end, it is assumed that the aggregate bit rate follows a Gaussian distribution.

Considering n multiplexed connections, the method determines the total bandwidth, c, that those n connections will need, such that

$$\text{Prob.\{instantaneous aggregate bit rate} > c\} < \varepsilon$$

where ε is a given value. The formula used to calculate the necessary aggregate bandwidth is as follows:

$$c = m + \alpha\sigma$$

where c = necessary aggregate bandwidth;
   m = total average bit rate;
   σ = total standard deviation.

Both the total average bit rate and the total standard deviation are calculated (and updated, when a new connection arrives) by adding the mean bit rate and standard deviation values of each connection. Several expressions have been suggested for determining the value of the parameter α; one of them is (see [3])

$$\alpha = \sqrt{-2 \cdot \ln(\varepsilon) - \ln(2\pi)}$$

The connection acceptance criterion is straightforward. The new VC is accepted if the bandwidth c required for all VCs including the one requesting admission is less than the available capacity of the VP or link, respectively. The Gaussian approximation is a very simple method. However, there are two weak points worth noting.

1. The method assumes a Gaussian distribution for the aggregate traffic; this only makes sense when there is a great number of connections present in the system or when the individual connections follow a Gaussian distribution.
2. No distinction is made (or taken into account) between the possibly different quality of service requirements of the individual connections, i.e. it is assumed that all connections have the same cell loss requirements.

1.5.2 Convolution method

The convolution method relies on knowledge of the distributions of the random variables representing the instantaneous rates of the individual connections. Instead of making an assumption about the distribution corresponding to the aggregate traffic (as with the Gaussian approximation method), the approach is to calculate it directly. Based on this result, the probability that the aggregated rate (including the new connection) exceeds the capacity limit of a VP or a link can be calculated and considered as a measure of the cell loss probability. The acceptance criterion then is not to exceed a certain prescribed cell loss probability.

In general, the convolution algorithm provides accurate results. The main problems with this algorithm are the availability of the distributions and the computing time necessary to calculate the distribution of the aggregate traffic's rate by convolution. Two solutions have been suggested for handling this last problem: one that stores the relevant distribution of already accepted connections and another that doesn't. In the first case, when there is a new connection set-up, only one convolution needs to be calculated and when there is one connection release, only one deconvolution will be needed. When the second solution is considered, a number of convolutions equal to the number of established connections has to be performed each time there is a new connection set-up and no deconvolutions are necessary in the case of a connection release.

Finally, it should be noted that in general, most CAC methods are either simplifications or adaptations of the two methods just described.

REFERENCES

[1] Saito H, *Teletraffic Technologies in ATM Networks*, Artech House, 1994
[2] Bensaou B, Guibert J and Roberts JW, *Fluid Queueing Models for a Superposition of ONIOFF Sources*, ITC Seminar, Morristown, NJ, 1990
[3] Onvural R O, *Asynchronous Transfer Mode Networks: Performance Issues*, Artech House, 1994
[4] Zhang H and Ferrari D, Rate controlled service disciplines, *J. High Speed Networks* **3**, 389–412, 1994
[5] Worster T, *Neural network based controllers for connection acceptance*, 2nd RACE Workshop on Traffic and Performance Aspects in IBCN, Aveiro, Portugal, January 1992
[6] Hiramatsu A, ATM communications network control by neural networks, IEEE Trans. Neural Networks, **1**, (1), 122–30, March 1990
[7] Castelli P, Cavallero E and Tonietti A, *Policing and call admission problems in ATM networks*, ITC-13, 847–852 1991.
[8] ITU-T (CCITT) recommendation
Q.2931 *B-ISDN Digital Subscriber Signalling (DSS2) User–Network Interface Layer 3 Specification for Basic Call/Connection Control*, 1993
[9] ATM Forum, *ATM User–Network Interface Specification, Version 3.1*, July 1994

1.6 USAGE PARAMETER CONTROL

According to ITU Rec. I.371 (see [1]), usage parameter control (UPC) "is the set of actions taken by the network to monitor and control traffic, in terms of traffic offered and validity of the ATM connection at the user access network". In other words, after a connection is accepted by the network using a connection admission control (CAC) procedure, it must be ensured that network resources are protected from malicious or unintentional misbehaviour of the source producing the call, which can affect the quality of service (QoS) of other calls already established in the network. This is the job of the UPC function, sometimes also called traffic policing and access control, situated at the access point to the network, i.e. the user Network interface (UNI). Typical locations for the UPC function include the entrances to a local switching node and a cross-connect.

The monitoring of traffic sources in a network is performed both by detecting violations of traffic parameters negotiated at call set-up and by taking action if violations occur. To this end, UPC is performed for each user VP/VC being used in the network and the following tasks are executed:

a)   the validity of each VP/VC's identifier is checked;

b)   the number of cell arrivals is counted for each connection established.

With this information, UPC can then compare the declared number of cell arrivals with the observed one. The non-agreement of the two values may lead to penalties being imposed on the traffic source observed. The two main actions that can be taken by the UPC at cell level on a misbehaving traffic source are:

- tagging of violating cells;
- discarding of violating cells.

In the first case, the UPC function can mark the violating cells as low priority cells by using the cell loss priority (CLP) bit existent in the header of ATM cells, as suggested by ITU Rec.I.371. This means that, when a situation of congestion occurs in the network, the marked cells will be lost first. An alternative way to tag cells could be to use a payload type field in the cell header (see [2]).

Each request of a user to establish a call in the network is initiated by the so-called traffic contract. This comprises the declaration by the user of certain traffic parameters that describe the characteristics of the call request. So far, the only traffic parameter standardised by ITU is peak cell rate. Other parameters have however been suggested to characterise traffic sources, such as:

Access to B-ISDN via Passive Optical Networks. Edited by U. Killat
© 1996 John Wiley & Sons Ltd

- average cell rate;
- burstiness;
- burst duration;
- sustainable cell rate (within the ATM Forum).

The problem with some of these parameters is that, although they contain more information about the traffic source that they represent, it is felt that normal users may not be able to be very precise about their sources' characteristics. In Chapter 1.6.3, more detail is given of the UPC mechanism proposed by ITU.

1.6.1 Algorithms for UPC

A policing algorithm should always be able to detect and quickly respond to any traffic violation in the network. It should also not take any action on traffic that conforms, i.e. traffic that doesn't violate the traffic contract negotiated at call set-up. Several policing algorithms have been proposed (see for example [3,4,8]) to regulate traffic flows in a network by controlling the sources' traffic parameters. In most cases, the controlled parameters are peak and mean cell rates. The algorithms can be divided into two main groups:

- window based mechanisms;
- leaky bucket (LB).

The first group considers algorithms in which fixed or variable time windows limit the number of cell arrivals, while the LB mechanism and its variants are based on a counter that is incremented whenever there is a cell arrival and decremented in the opposing case. Examples of window algorithms are schemes like jumping window (JW), triggered jumping moving window (TJMW) and the exponentially weighed moving average (EWMA). In the JW mechanism, fixed and consecutive intervals with X time slots are observed; they allow a maximum of Y cell arrivals. If more than Y cell arrivals are observed in a time window, the first Y cells are allowed to pass and the others are lost (or marked). When each time window ends, the counters used to evaluate the number cell arrivals and time slots are reset and a new time window begins. TJMW is a variation of this mechanism; it considers time windows that are not necessarily consecutive and are started by a new cell arrival. Although similar to the JW, the EWMA scheme doesn't consider a fixed maximum number of cell arrivals in each time window. The number of cell arrivals that can be accepted in each time window is an exponentially weighed sum of the number of accepted cells in the previous time window and the mean number of cells.

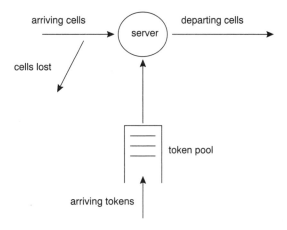

Figure 1.6.1
The leaky bucket mechanism

The leaky bucket algorithm (see Figure 1.6.1) is probably the most well known algorithm that can help UPC in judging a violation in the connection set-up phase and a number of variants can be found in the literature (see [3,9,10,11]); the method is thought to be simple and flexible. The idea behind LB is that, when a cell arrives in the network, it must get a token from the token pool. If there are any tokens available in the token pool on cell arrival, the cell takes one token and the token pool size is decreased by one. The tokens are generated at a constant rate and the token pool is like a virtual finite queue. Cell arrivals are discarded (or tagged) when the token pool is empty. The algorithm can be schematically described in its simplest form by the following:

1. increment a counter C by 1 every T seconds up to M if there are no cell arrivals; $(0 \leq C \leq M)$

2. decrement the counter C by 1 for each transmitted cell;

3. drop (or tag) cells when $C = 0$.

M represents the maximum burst size, while $1/T$ is the peak cell rate of the traffic source being observed.

Both VPs and VCs can be policed by using the LB scheme. When VP policing is performed, the only measure of interest is the aggregate bit rate offered to the VPs, which can usually be easily policed; therefore, the contents of each VC multiplexed in a VP and their source characteristics are ignored for policing purposes. However, when policing is executed on a VC basis, it becomes more complicated to control the sources' traffic parameters and, as in this case, the sources have to be characterised by at least their peak and mean cell rates.

Several authors have addressed the problem of comparing the performance of the different policing mechanisms proposed (see for example [3,12,13]). Of the window mechanisms, the JW scheme is considered to be the simplest but also the one that gives the worst performance. The LB algorithm is usually acknowledged as giving better results than window based schemes. LB is also the policing mechanism recommended by the ITU and will be described in more detail in Chapter 1.6.3.

It has been demonstrated that sources' traffic streams are better controlled when violating cells are discarded, as opposed to tagging them. However, this solution implies stringent requirements for the violating probabilities of well behaved traffic sources. On the other hand, when marking of violating cells takes place, it is not possible to distinguish between marked cells of sources that are really misbehaving and cells of well behaved sources that have been mistakenly marked by the policing mechanism used.

1.6.2   CDV and UPC

In an ATM network, the cell delay experienced by cells of the same connection can vary. This is called cell delay variation (CDV) and is caused mainly by buffering at ATM switching nodes. Since CDV can change the traffic characteristics of sources in the network, the bandwidth required by those sources can also change. Therefore, UPC should take CDV into account when deciding whether cell arrivals are violating the respective traffic contract negotiated at call set-up. In this case, the maximum allowed CDV should be an extra parameter to be included in the traffic contract.

When UPC takes CDV into account, it introduces tolerances in the amount of traffic from a source that conforms to the corresponding traffic contract. It is thus possible that some cells that are really violating the traffic contract are not discarded. One way to compensate for this effect is for the network to introduce shaping just before UPC is performed. This reduces the CDV generated in a network ([2]). Shaping is used to smooth the clumping of cells from a traffic source by using buffers in which cells are put before they enter the network. The buffers are read at a rate determined by the shaping process. Although this process can reduce the clumping, it does so at the expense of delaying the traffic generated by the source; thus, it is a process that can only be tolerated by non-delay-sensitive traffic.

1.6.3   ATM Forum *vs.* ITU

Recently, both the ITU and the ATM Forum have been addressing the subject of policing. The policing algorithm used by the ITU (see [1]), known as the continuous-state leaky bucket algorithm (CSLB), uses two parameters: the peak emission interval T (which is the inverse of the peak cell rate of an ATM connection) and the CDV tolerance tau. Let

C   = value of the leaky bucket counter
aux_var = auxiliary variable
lct   = last compliance time
t_arrival = time of a cell arrival.

The algorithm is then as follows:

```
record first cell arrival in t_arrival;
C = 0;
lct = t_arrival;
while there are cell arrivals do
    begin
        record a cell arrival in t_arrival;
        aux_var = C - (t_arrival - lct);
        if (aux_var < 0) then
            begin
                aux_var = 0;
                C = aux_var + T;
                lct = t_arrival;
                arrival in t_arrival is a compliant cell;
            end
        else begin
            if (aux_var > tau) then
                    arrival in t_arrival is a non-compliant cell
            else begin
                    C = aux_var + T;
                    lct = t_arrival;
                    arrival in t_arrival is a compliant cell;
                end;
            end;
end;
```

The decision of considering a cell as non-compliant (or non-conforming) implies one of two actions, as referred to before: tagging or discarding. In the first case, the tagging of a cell consists in setting the CLP bit to 1, thus transforming it into a low priority cell. Two scenarios are possible (see [1]) for deciding on whether to allow a cell (be it low- or high-priority) to pass, to be marked or discarded.

a) *no cell tagging* With this strategy, UPC starts by checking if the CLP $= 0$ stream (representing high-priority traffic) is conforming (i.e., is respecting the agreed traffic contract). Any cells that are

found to be non-conforming will be discarded. A second conformance test is then performed on the aggregate traffic stream (i.e.,
the $CLP = 0 + 1$ stream). The aggregate traffic stream is now the
sum of the conformant $CLP = 0$ traffic (after the first conformance
test) and the $CLP = 1$ traffic (i.e., low-priority traffic). The second
conformance test will again discard any cells (irrespective of their
CLP bit value) that do not respect the agreed traffic contract.

b) *cell tagging* Two conformance tests are performed in this case. The
    first will again check whether the $CLP = 0$ stream is respecting the
    traffic contract of the corresponding traffic source. Cells that violate
    the traffic contract will be tagged (i.e., their CLP bit will be changed
    to 1). After this test, the $CLP = 1$ will consist of the original $CLP =
    1$ stream and the tagged $CLP = 0$ cells. The second conformance
    test is once more applied to the aggregate traffic, which consists
    of the $CLP = 0$ and $CLP = 1$ traffic streams' sum. Any cells (be
    they low-priority — $CLP = 1$, or high-priority — $CLP = 0$) that are
    found to violate the traffic contract will be discarded.

The conformance tests are executed by making use of the chosen algorithm, which in this case is CSLB.

The generic cell rate algorithm (GCRA) proposed by the ATM Forum
is basically a CSLB algorithm ([6]). However, this organisation is also
considering UPC algorithms with three components:

- a peak rate controller for the aggregate cell stream ($CLP = 0 + 1$);
- maximum sustainable rate controllers for both low and high-priority
  cell streams ($CLP = 1$ and $CLP = 0$, respectively).

Moreover, the ATM Forum is also considering the possibility of using
the available bit rate (ABR) as another possible parameter for characterising traffic sources.

Finally, another UPC algorithm (based on the LB algorithm) is being
considered by the ATM Forum (see [5,7]): the dual leaky bucket algorithm. With this scheme, two leaky bucket algorithms are executed in
parallel. The two LBs will be described by the parameters (T_1, M_1) and
(T_2, M_2), in a way analogous to that of the description of LB given in
Chapter 1.6.1. This means that $1/T_1$ and $1/T_2$ are the leak rates and M_1
and M_2 are the bucket sizes. Each cell arrival will go through both leaky
buckets and it will be considered to be conforming if the following
condition holds:

$$C_1 \leq M_1 \quad \text{and} \quad C_2 \leq M_2,$$

where C_1 and C_2 are the leaky buckets' counters (see Chapter 1.6.1).

If $T_1 < T_2$, then the first leaky bucket is known as the peak rate leaky
bucket and the second is named the sustainable rate leaky bucket,

from which it follows that $1/T_1$ is called the peak rate, $1/T_2$ is the sustainable rate and M_2 is the burst tolerance. It is felt that a double leaky bucket mechanism will be more efficient in terms of policing traffic sources than a simple leaky bucket algorithm. Note also that the policing schemes proposed by the ITU (see [1]) consider a process with two leaky buckets (one for the CLP $= 0$ stream, i.e. the high-priority traffic, and another for the CLP $= 0 + 1$ stream, i.e. the aggregate traffic).

REFERENCES

[1]   ITU-T (CCITT) recommendation I.371 *Traffic Control and Congestion Control in B-ISDN*, 1993
[2]   Saito H, *Teletraffic Technologies in ATM Networks*, Artech House, 1994
[3]   Buttó M, Cavallero E and Tonietti A, Effectiveness of the Leaky Bucket Policing Mechanism in ATM Networks, IEEE Selected Areas in Communications, **9**, (3), 335–42, 1991
[4]   Cuthbert L G and Sapanel J-C, *ATM, The Broadband Telecommunications Solution*, IEE Telecom. Series 29, 1993
[5]   *ATM Forum: Leaky Bucket Based UPC Algorithms*, Doc No 92-256, November 1992
[6]   *ATM Forum: Performance Analysis of GCRA for CBR Sources*, Doc No 94-0182, March 1994
[7]   Mark B L and Ramamurthy G, *UPC Based Traffic Descriptors for ATM: How to determine, interpret and use them*, Proceedings of the ATM Traffic Expert Symposium, Basel, Switzerland, April 1995
[8]   Onvural R O, *Asynchronous Transfer Mode Networks: Performance Issues*, Artech House, 1994
[9]   Ahmadi H R, Guérin R and Sohraby K, *Analysis of Leaky Bucket Access Control Mechanisms with Batch Arrival Process*, GLOBECOM, 1990
[10] Castelli P, Cavallero E and Tonietti A, *Policing and Call Admission Problems in ATM Networks*, ITC-13, 1991, 847–52
[11] Chao H J, *Design of Leaky Bucket Access Control Schemes in ATM Networks*, ICC '91, 1991, 180–87
[12] Rathgeb E P, *Modelling and Performance Comparison of Policing Mechanisms for ATM Networks*, IEEE Journal on Selected Areas in Communications, **9**, (3), April 1991, 325–34
[13] Kim Y H, Shin B C and Un C K, *Performance Analysis of Leaky-Bucket Bandwidth Enforcement Strategy for Bursty Traffics in an ATM Network*, Computer Networks and ISDN Systems, **25**, (3), 1992, 295–303

1.7   INTRODUCTION SCENARIOS

ATM technology has found its champions amongst people from the telecom industry and from the computer industry. As a result, it is not obvious what the most promising introduction scenarios might be. A very pragmatic viewpoint might be: Whatever the potential benefits of ATM, it also has to support the existing world. This statement might mean a number of things.

- Existing networks and their native terminals are coupled via an ATM network: the backbone approach.

- An existing network and an ATM network are coupled. Terminals on both sides designed for the same service can talk to each other: the interconnection approach.

- An existing network supporting a given set of terminals and applications is replaced by an ATM network without adversely influencing the applications: the substitution approach.

These different approaches are reflected in the concepts presented in the following sections. Another way of introducing ATM would be to offer new applications on existing networks by upgrading them to reach ATM compatibility. An exercise of this sort, namely deploying ATM services on passive optical networks, is described in detail in Section 3 of this book.

1.7.1   B-ISDN in support of ISDN

The backbone approach is already becoming a reality in two different areas.

- The telecommunication network operators plan to install ATM cross-connects for routing higher-order multiplexes of their signals and for more efficient management of their network resources. In this way, ISDN networks will be concatenated via a B-ISDN network. The ISDN is not aware of the nature of the long haul transport facilities and the ATM network is not aware of the nature of the data streams it transports.

- The backbone approach can also be seen in collapsed systems. Public or private ISDN switches are constructed from an ATM kernel which is then equipped with ISDN interfaces. The philosophy behind this concept is to benefit from the low-cost high-throughput switching technology and to have an internal interface that allows for an easy system migration towards new ATM type interfaces.

Access to B-ISDN via Passive Optical Networks. Edited by U. Killat
© 1996 John Wiley & Sons Ltd

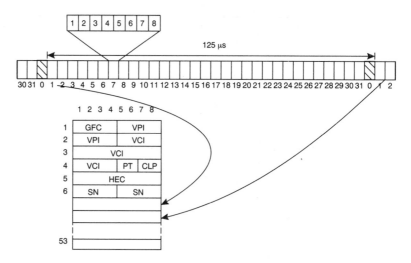

Figure 1.7.1
Mapping of PCM slots onto cells

In any case, at the interface between the ISDN and the ATM network an interworking unit (IWU) is necessary which maps pulse code modulation (PCM) slots into cells and vice versa (Figure 1.7.1).

In its simplest form the backbone approach can be seen as an encapsulation technique. The IWU does not intervene in the content of the time slots but transparently transmits them in the format of cells across the ATM network.

When interconnection is considered, the situation changes. Now the IWU intervenes up to and including the Layer 3 signalling protocol which happens to be Q.931 on the ISDN side and Q.2931 on the B-ISDN side; see Figure 1.7.2 and [1].

The translation of service primitives in the IWU allows two ISDN terminals on both networks to get connected to each other. The IWU in this case has to be equipped with the corresponding UNI interfaces. The recommendation I.580 [1] also describes an alternative where the IWU supports NNI signalling. It is also stated that mappings such as the one described in Figure 1.7.2 require further study.

A frequently discussed problem is the synchronisation between the terminals involved, whereby it is assumed that ISDN terminals on an ISDN network are synchronised to the network clock of that ISDN network. The problem is further discussed here with reference to Figure 1.7.3.

If the ISDN networks 1, 2 and 3 have plesiochronous clocks, slips will occur in almost the same manner as today and this will be tolerated according to the respective CCITT recommendations. If the clocks involved are asynchronous to each other, one clock can be selected to

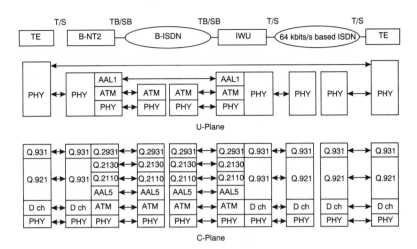

Figure 1.7.2
Interworking scenario based on IWU equipped with UNI interfaces

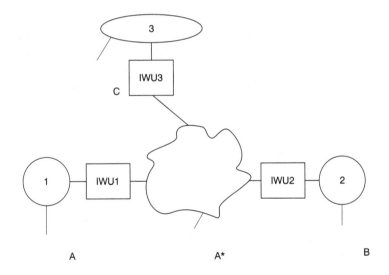

Figure 1.7.3
Network configuration used to illustrate the problem of network synchronisation
(see text)

become the master, for example the clock of network 1. In this case
the clock of the ATM network has to be slaved to clock 1 in IWU 1,
whereas the clocks of networks 2 and 3 are slaved to the clock of the
ATM network in their respective IWUs. The clock recovery function of
AAL-1 would not be of any great help in the situation of Figure 1.7.3: It
allows a receiver to synchronise with the clock of the sender. However,
an ISDN terminal can support two connections. Therefore, if terminal
B could maintain two connections (to terminals A and C) it could not

synchronise itself simultaneously with both of them. Therefore AAL-1 is only helpful in a situation where only network 1 and the ATM network exist. Then terminal A* can indeed profit from AAL-1 and regenerate the clock of terminal A. Similarly, if networks 1 and 2 were concatenated via the ATM network, IWU 2 could regenerate by the help of AAL-1 the (master) clock of network 1 and thereby synchronise network 2 with network 1. In this case, the ATM network can run asynchronously to both of the other networks.

1.7.2   ATM technology in support of LANs

LANs, as they are defined according to IEEE 802, are based on shared media interconnects. At the medium access control (MAC) layer these networks operate connectionlessly. Each host attached to a LAN has a globally unique MAC address which is six bytes in length and has a flat address space. An ATM network can be used to interconnect two LANs to give the appearance of a geographically widespread LAN (Figure 1.7.4a).

In this case the backbone function of an ATM network is exploited. A permanent virtual circuit is used to interconnect both parts of the LAN. To realise this scenario an interworking unit (IWU) is necessary which performs the segmentation and reassembly of MAC frames to/from ATM cells. In ITU-T terms this is an example of an indirect provision of

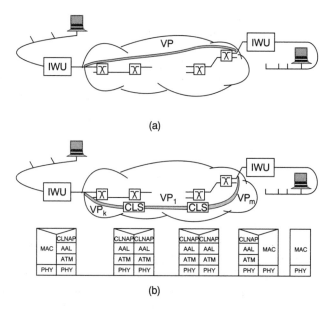

Figure 1.7.4
Backbone function of ATM network. (a) indirect provision of connectionless service; (b) direct provision of connectionless service

a connectionless service: the ATM network is not aware of connection-less communication and does not intervene in connectionless protocols. In contrast, in Figure 1.7.4b a connectionless overlay network has been created: there are special nodes inside the ATM network, so-called connectionless servers (CLSs), which are peers of the connectionless network access protocol (CLNAP) which is implemented in IWUs or in workstations directly connected to the ATM network. This scenario is referred to as the direct provision of a connectionless service. The principles of both scenarios seem to be straightforward, but they result in some intricate problems when details of the implementation are considered. This will become obvious in the following subchapters.

1.7.2.1   *Indirect service provision*

In Figure 1.7.4 the coupling of two LANs via remote bridges (the IWUs) has been described. If the interconnection of a couple of LANs (and/or metropolitan area networks (MANs)) is considered, the IWUs have to select a route across the ATM network in order to reach the destination LAN. To do this, different concepts have been developed of which "LAN emulation" (ATM Forum) and "IP over ATM" (Internet Engineering Task Force [2]) are the most popular ones.

In the LAN emulation approach the ATM layer is seen as just another MAC layer. However, from the perspective of a classical LAN this new MAC layer has two deficiencies:

- it has no inherent broadcast capabilities
- it introduces a new addressing scheme.

LAN emulation (LE) is a set of services which have to be implemented in systems (hosts or IWUs) which are directly connected to the ATM network (see Figure 1.7.5). For the realisation of these services a client-server architecture has been developed: a client can invoke the service

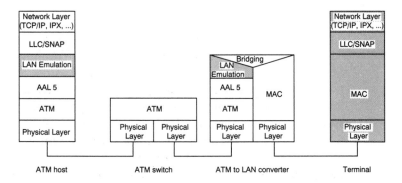

Figure 1.7.5
Protocol stacks for LAN emulation

offered by a set of servers. The most important service is to perform address resolution, i.e. to present the ATM address that goes along with the destination MAC address. When a MAC DATA request is presented to an LE client he has to check whether the ATM address corresponding to the MAC address is already in the cache. If this is the case, the next check will determine whether an ATM connection already exists to this destination. If the ATM address is not known it can be retrieved with the help of the LE servers. Similarly, a server will be addressed if multicast or broadcast transmission is required. Once the ATM address has been obtained, a connection set-up will provide the desired connectivity.

The LAN emulation concept has the virtue that above the MAC layer everything remains unchanged. Therefore this approach is also suitable when an existing LAN has to be substituted by a more powerful ATM LAN based on ATM switches. But there are also drawbacks. The concept of different servers, the details of which have been concealed from the reader, indicates a high degree of complexity. In particular if servers are implemented in a distributed manner in a multi-vendor environment, seamless integration of a network seems important. LAN emulation is LAN dependent and only applies to one type of LAN. Up to now it does not include inter-operability of Ethernet and Token Ring for example. This drawback is avoided in the "IP over ATM" approach which in turn is network protocol (IP) dependent.

IP over ATM tries to strictly preserve the mechanisms and protocols existing in classical IP networks [2], in particular the address resolution protocol (ARP). In order for a host to send data to another host he needs to know the MAC address of the destination or of the next router. In an IP over ATM environment, it is necessary to derive the ATM address (of the next router) from the IP address of the destination. Again, it has been proposed to install a server, the IP-ATM-ARP server, which can be reached via a globally known VC (or ATM address). Unlike the situation of LAN emulation, no broadcast facilities are needed at the IP level.

The interaction between the hosts/routers and the server can be based on a simple query/response protocol. The query specifies the destination IP address, the response an ATM address. If the target is to leave the structure of the IP network as it is, the ATM address is always that of the next router: the IP network layer architecture prevents end systems from establishing direct communications when they belong to different logical internet subnets. Thus a datagram is forwarded on a hop-by-hop basis from one router to the next. This situation would also hold in an IP over ATM implementation even though there might exist a direct link between the router of the source network and that of the destination network! If classical IP is implemented on top of

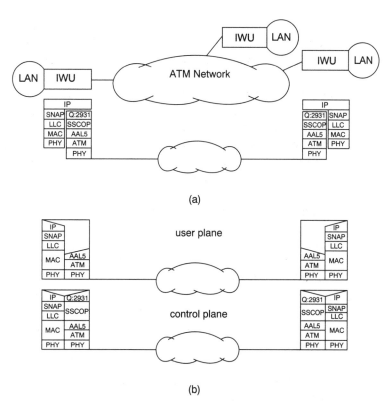

Figure 1.7.6
IP over ATM: Router protocol stacks. (a) classical IP; (b) integrated architecture

ATM interconnects the latter are only seen as a data link layer. This view is also reflected in Figure 1.7.6a. Address resolution is followed by connection establishment: if a connection to the destination already exists, the corresponding VCI will be used. Otherwise data transfer will be preceded by a connection set-up phase.

The classical IP approach does not fully use the capabilities of the underlying ATM network. What is needed is

- the possibility of directly interconnecting any two hosts/routers which are hooked onto the ATM network
- getting rid of the tedious and time-consuming AR protocols.

As has been explained in [3] the most important prerequisite for such an approach would be to have a one-to-one algorithmic mapping between IP and ATM addresses. As a consequence, address resolution would become superfluous and IP subnetworks, as well as the ATM network, would form one network for which one integrated routing algorithm and routing protocol would apply. All multiplexing, for

INTRODUCTION SCENARIOS

example for different network protocols or even transport connections, can be mapped onto virtual channel connections. Thus the IP header becomes superfluous inside the ATM network. The respective protocol stacks for a router between a classical LAN and the ATM network are depicted in Figure 1.7.6b, the control plane is also shown to indicate the different level of involvement during the signalling and data transfer phases. The question of whether such an approach has to be ascribed to indirect service provision is debatable: the criterion is the protocol level up to which the network inspects the data. The criterion itself is clear. What is not unambiguously being defined is "the network". If the routers are not considered as part of the network, the paradigm of indirect service provision applies; otherwise the depicted scenario is also an example of direct service provision.

1.7.2.2 Direct service provision

In the ITU-T terms, direct service provision relates to connectionless servers (CLSs) forming an overlay network above the ATM network. They receive datagrams which, in the ITU-T world, are equipped with E.164 addresses. The CLS inspects the address of the datagram and determines an outgoing VPI which leads directly or via subsequent CLSs to the destination. The approach is particularly appealing when the connectionless service is based on semi-permanent VPs. In the indirect service provision scenario, each IWU would need a VP (and a certain bandwidth reserved for it) for every potential destination IWU. In the direct approach, only one VP is reserved for communication with the CLS and thereby the reservation of bandwidth resources is fairly limited.

As described above, the basic function of the CLS is its routing function. There are two basic approaches to how this function can be executed and they also represent a trade-off in terms of network performance and cost. The two different principles are sketched in Figure 1.7.7. In the first case a complete datagram is first reassembled in the CLS and then forwarded to the outgoing VP. The obvious advantage of this "frame mode" is that only complete and uncorrupted messages are released from a node. This has to be paid for by a substantial amount of reassembling buffer. This reassembling buffer is superfluous in the second approach where the CLS works in the "cell mode". A hardware supported address to VP resolution scheme yields the outgoing VP in only a few (e.g. 6) cell times [4] after the BOM cell containing that address has been received. Thus forwarding of cells on the outgoing VP can start as soon as that VP has been made available without waiting for the arrival of the remaining cells belonging to the same datagram.

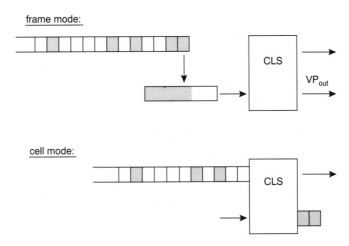

Figure 1.7.7
Alternative modes of operation for a connectionless server

Owing to the random nature of the arrival of datagrams, prevention of congestion is a critical issue for a CLS. For the operation of a CLS the AAL-3/4 protocol has been proposed and datagrams may arrive cell-interleaved. Congestion may occur when more than one arriving datagram has to be forwarded to the same outgoing VP. Then datagrams will have to be buffered and may or may not find sufficient buffer space.

This situation is similar to the classical problem of cell buffering in an output buffered system as discussed with reference to Figure 1.3.3. A cell loss will invalidate the message the cell belongs to. At a cell loss rate of 5×10^{-3}, every cell loss may hit a different datagram resulting in the worst case of a zero throughput in terms of datagrams. One way of mitigating the problem is datagram acceptance control (DAC) [5]. The CLS, upon reception of a BOM cell, derives an estimate of whether the incoming datagram will create a buffer overflow. If a buffer overflow is predicted, all cells of that datagram are discarded. Otherwise, the datagram is said to be accepted and its cells are put into the buffer. By this procedure datagram losses can be reduced by an order of magnitude. The buffers needed for datagrams contending for the same outgoing VP are by an order of magnitude smaller than the reassembling buffers needed in the frame mode. This corroborates the strength of the cell mode.

1.7.3 ATM technology in support of the frame relay service

The frame relay service has become very popular in recent years because of its simplicity and relatively appealing performance. It offers a connection oriented service and statistical multiplexing over lines from 56 kbit/s up to 45 Mbit/s. User frames have a size of up to 8 Kbytes and are protected by a CRC-16. Error recovery by means of retransmissions is left to the user of the service.

The frame relay service can be supported by an ATM network on the basis of a frame relay service specific convergence sublayer (FR-SSCS) which resides on the CPCS of AAL-5. The protocol data units of the FR-SSCS look very much like the PDUs specified in recommendation Q.922 for the frame relay service. Again, on the basis of suitable interworking functions, the ATM network can act as a backbone interconnecting transparently two or more frame relay islands. Similarly to the connectionless service, for the frame relay service, servers can be thought of which are connected to the ATM network or which form an integral part of it. In the latter case the ATM network would intervene in the frame relay supporting protocol stack, up to and including the FR-SSCS. For more details the interested reader is referred to recommendation I.555 [1].

REFERENCES

[1] ITU-T (CCITT) recommendation
     I.555 *Interworking between Frame Relaying Networks and B-ISDN*, 1993
     I.580 *General Arrangements for interworking between B-ISDN and 64 kbit/s-based ISDN*, 1993
     Q.2931 *B-ISDN Digital Subscriber Signalling (DSS2) User–Network Interface Layer 3 Specification for Basic Call/Connection Control*, 1993
[2] Laubach M, *Classical IP and ARP over ATM*, RFC 1577, 1994
[3] Lian F C, Perkins D, *Beyond Classical IP-integrated IP and ATM Architecture Overview*, ATM Forum Contribution 94-0935 1994
[4] Henkel V, Frankenfeld C, Klapdohr K, Menzer K and Schmidt A, *A High Performance ATM/DQDB Cell to Slot Interworking Unit*, Proc. Broadband Islands '94: Connecting with the End-User, W. Bauerfeld, O. Spaniol and F. Williams (eds), 453–56, North-Holland: Amsterdam, 1994
[5] Vogt R and Killat U, *Performance Evaluation of Datagram Acceptance Control in DQDB to ATM Interworking Unit*, Proc. International Switching Symposium, Berlin, 1995, paper B1.7

Section 2

PASSIVE OPTICAL TREE NETWORKS

2.1 INTRODUCTION

A passive optical network (PON) is an optical network without any additional electronic or opto-electronic devices. A PON may thus consist of optical fibre, optical splitters and combiners, directional couplers, lenses, gratings, optical filters, phase masks etc. PONs have several advantages over networks with active components: they do not require an electric power supply and are consequently not sensitive to power failures; PONs are not EMI sensitive; and, what is important for the cost of operation: PONs are highly reliable and require no maintenance, because there is no degradation as there would be with active components. The optical fibre used will generally be a single-mode fibre for broadband networks. One important aspect of a multi-user PON is its physical topology, i.e. the way in which fibre parts are connected to form a network. For the application as a subscriber access network the logical structure requires one bidirectional point-to-point connection between the optical line termination (OLT) and each optical network unit (ONU). This can be accomplished by different physical topology types.

2.1.1 Network topology

For the above-mentioned application, the following physical topologies could be used, see also Figure 2.2.1.

- Bus: this is one fibre consecutively running from one station to the next. Each station is connected to the fibre via a directional coupler.
- Ring: same as bus, but with the open ends being connected.
- Star: this may be implemented in either of two ways.
    - Using a transmissive coupler which employs two dedicated fibres for each ONU, one for each direction (transmit/receive).
    - Using a reflective coupler whereby each ONU is connected to the coupler by only one fibre for both directions of transmission.
    All these fibres are connected to a passive optical star coupler that broadcasts a superposition of all optical input signals to all of the outputs. This leads to a logical bus structure.

It must be mentioned here that the term "star" is not used consistently in the literature. While [1] uses the above definition, which is appropriate for passive networks, there are also examples of a different definition: [2,3,4] define a star as a network with one dedicated fibre from each ONU to the OLT.

Access to B-ISDN via Passive Optical Networks. Edited by U. Killat
© 1996 John Wiley & Sons Ltd

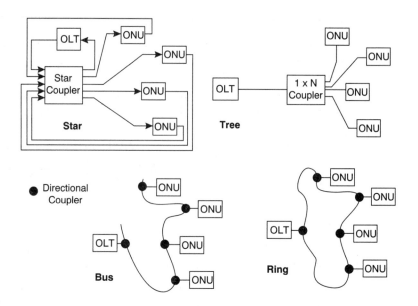

Figure 2.1.1
Passive optical network topologies: star, tree, bus and ring. OLT, optical line termination ONU, optical network unit

- Tree: a tree starts with one fibre from the OLT and has several fibre branches to connect the ONUs to the stem.

The tree topology is often named "multiple star" ("double star" if there is only one branching level in the network) [2,3,4].

These topologies, which allow sharing of the transmission medium amongst many subscribers should be used for residential and medium-sized business customers who will not really need the huge potential bandwidth of a broadband network, but nevertheless must be offered a cost-efficient network access. The star topology is physically not very different from a tree and has the logical properties of a bus. A *passive* ring is not easily made, because the signal must be removed selectively from the ring after one circulation which is not possible with passive components. The bus structure does not present this problem, but for a bus as well as for a ring the optical power splitting ratio is a problem: either couplers with the same ratio are used for all stations, which leads to very low received power at the farthest stations (known as near–far problem), or couplers with tuned ratio are used to give the same power to each of the stations. Both approaches are neither very flexible nor cheap. Finally, the tree topology exhibits none of the above-mentioned problems and is thus the preferred choice for local loop applications.

Once the tree topology is chosen, a variety of problems must be treated which are covered in more depth in the following chapters.

Chapter 2.2 will examine the technological questions concerning PONs: optical components to be used, optical power budget, transceiver requirements. Chapter 2.3 is on the topic of multiple access methods: for the upstream[†] traffic a way must be found to communicate on the stem fibre shared by all the subscribers without causing collisions. Chapter 2.4 covers the multiplexing for the upstream and downstream directions: this can be accomplished on one dedicated fibre tree for each direction, but there are also various methods for transmitting bidirectionally on one fibre. Finally, Chapter 2.5 will give an overview of the possibilities and installed base of PONs in an ISDN environment.

REFERENCES

[1]   Green P E, *Fiber Optic Networks*, Prentice Hall: Englewood Cliffs, 1993
[2]   Kashima N, *Optical Transmission for the Subscriber Loop*, Artech House: Boston, 1993
[3]   Reed D P, *Residential Fiber Optic Networks*, Artech House: Boston, 1992
[4]   Sporleder F *et al*, *Optische Übertragungstechnik für flächendeckende Teilnehmeranschlüsse*, volume 46 of *Der Fernmeldeingenieur*. Verlag für Wissenschaft und Leben Georg Heidecker GmbH: Erlangen, 4th edn, April 1992.

[†] i.e. from the subscriber ends to the local exchange

2.2 TECHNOLOGY

2.2.1 Passive optical components

Passive optical components are used in the optical distribution network (ODN) and at the optical interfaces connecting the termination equipment with the ODN. The ODN takes care of the connection between the optical network units (ONUs) at the subscriber side and the optical line termination (OLT) at the network, or exchange, side. In the case of an ODN with a tree topology, a major task of the ODN is splitting and combining the optical signal flows between OLT and ONUs. The main components of the ODN are therefore the optical fibre and the optical couplers, having respectively the transport and the splitting/combining functions. Wavelength division multiplexers (WDMs) are wavelength dependent couplers. These are applied when the network makes use of multiple optical wavelengths. The physical connections of ONU and OLT to the ODN will be made with optical connectors. In the ODN, spliced fibre connections are used to make the permanent fibre interconnections.

2.2.1.1 Fibres

Fibres can be generally divided into two types: single-mode fibres and multimode fibres. The term multimode illustrates the existence of a large number of electromagnetic field modes that propagate through the fibre. The number of modes depends on the ratio between the fibre core diameter and the optical wavelength. When the core diameter is small, about 10 μm or smaller, only one mode, the fundamental mode, of the regular applied optical wavelengths can propagate through the fibre. Multimode fibre has a much larger core diameter of 50 μm to 100 μm.

For both fibres, the light is kept in the core of the fibre by applying different refractive indices for the core and the material around the core, called the cladding. The main consequential difference between single and multimode fibre is the dispersion. This is the technical term for pulse broadening. Compared to single-mode, multimode has a much higher dispersion, due to the existence of the different modes. Dispersion results in a signal deterioration in the time domain. Another fibre characteristic is the maximum fibre modulation bandwidth. This is a frequency domain property, directly related to dispersion. Multimode fibre is limited in bandwidth and is not suitable for high bit rate systems with network lengths regularly used in PON systems. For evolutionary reasons, this fibre is also not applied for low bit rate PONs. A single-mode fibre PON allows for future system enhancements to higher bit rates or different modulation techniques. Since multimode fibre is not used for access PONs, no further attention will be paid to this type.

Access to B-ISDN via Passive Optical Networks. Edited by U. Killat
© 1996 John Wiley & Sons Ltd

Table 2.2.1
Standard single-mode fibre properties according to G.652

wavelength (nm)	max. chromatic dispersion (ps/(nm.km))	min./max. attenuation (dB/km)
1285–1330	3.5	0.3–1.0
1270–1340	6.0	0.3–1.0
1550	20.0	0.15–0.5

There are, in general, three types of single-mode fibre [1]:

- standard single-mode (SSM) fibre
- dispersion shifted single-mode (DSSM) fibre
- dispersion flattened single-mode (DFSM) fibre

SSM fibre has almost zero dispersion at 1300 nm. But at this wavelength, fibres have a relatively high attenuation compared to the 1550 nm window. DSSM fibre has a zero dispersion at 1550 nm, and a relatively high dispersion in the 1300 nm window. Owing to the low attenuation and zero dispersion at 1550 nm, this fibre is ideally suited for systems using that wavelength. DFSM fibre has a low dispersion in both the 1300 and 1550 nm windows, making it a very attractive fibre in this respect. However, the cost of this fibre type is higher than that of the SSM fibre. Moreover, splicing DFSM fibre to the most commonly used SSM fused optical coupler is more difficult than splicing SSM fibre to such coupler. This makes DFSM fibre unattractive for tree PONs. It can be concluded that, in general, SSM fibre is the most suitable type for access tree PONs. Specification of this fibre type is given in ITU Recommendation G.652. Table 2.2.1 gives the standardised properties of this fibre.

2.2.1.2 Couplers

Couplers are reciprocal optical devices which can be used as a splitter and combiner. Two technologies are of primary interest: fused couplers and planar couplers [2]. Fused couplers are produced by fusing a number of fibres together and stretching the fibres over the length which is fused. The stretching decreases the diameter of the fibre, which causes a broadening of the field and coupling between the fields in the fibres to be coupled. A coupler with two input and two output ports is depicted in Figure 2.2.1. A coupler with a higher port count is made by twisting the fibres before fusing and stretching.

Fused couplers are available in a large variety of input and output port configurations. These couplers can be made achromatic, or suited to both the 1300 nm and 1550 nm window. The coupling uniformity is not very good. The coupling loss of fused couplers is approximately 3.3 dB per factor 2 splitting. The non-uniformity is evidenced by the loss difference in the output branches of the same coupler, which can be

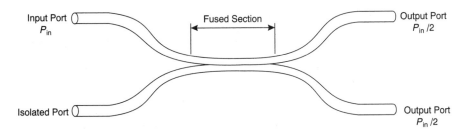

Figure 2.2.1
Optical coupler with two input and two output ports

maximally about 1 dB, resulting in coupling losses between 2.8 dB and 3.8 dB per factor 2 splitting. Reflection of fused couplers is very good: −40 dB can be realised. Directivity, or the relation between coupling in the wanted direction to coupling in the wrong direction, is also very good with fused couplers. A value of about 50 dB can be obtained.

Planar couplers are built on the substrate of an optical guiding material, having integrated optical waveguides. The substrate and waveguide materials have different refractive indices. These coupler types have an excellent uniformity. As this is a new technique that is still under development, not all input and output port configurations are commercially available. Moreover, planar devices are still quite expensive. These drawbacks make this type of coupler less suitable for PONs.

2.2.1.3 Wavelength division multiplexers

WDMs are wavelength dependent couplers. They are intended for splitting and combining signals with different optical wavelengths. Wavelength isolation is the most important parameter of a WDM. This parameter describes the ability to separate two signals with different wavelengths. Mutual deterioration of both signals will occur in the receiver when the isolation is not good enough. A regular application of these devices is in the case of a PON using two optical windows. A typical example is to use the 1300 nm window for interactive data traffic, combining it with TV distribution at 1550 nm over the same PON.

There are three types of WDM: fused, filter and mirror WDMs [2]. For application in PONs where only the 1300 nm and 1550 nm windows have to be (de)multiplexed, the fused type is the most suitable because of its relatively low costs. The mechanical construction is in principle the same as for the optical coupler. Wavelength dependency is obtained by choosing certain mechanical dimensions for the coupler. The main difference in performance is that the standard fused WDMs have a bad isolation, about 20 dB, compared with 40 dB for the other two types. However, high-grade fused WDMs can also achieve isolation values of about 40 dB when both wavelengths are not too close to each other.

WDMs are not usually used in the ODN but in the optical terminating equipment. In that way, the ODN is wavelength transparent and can be used for carrying different wavelengths. For isolation between the 1300 nm and the 1550 nm windows, which is the most common situation in PONs at the moment, high-grade fused WDMs are the best choice. Some indicative properties are: insertion loss 0.5–1.0 dB, reflection −40 dB and isolation 40 dB.

2.2.1.4 *Splices*

Because of the finite length of fibres as fabricated and delivered by the manufacturers, there is a need for durable, permanent fibre joints which can be made in the field. Such a joint is called a splice [2]. Splices are also used to connect the fibre pigtails of couplers to the ODN fibres. Two types of splice exist: mechanical and fusion splices.

Mechanical splicing is based on a V-groove mount fixing the two fibres to be spliced. Index matching liquid is necessary to provide a good coupling between the fibres. This type of splicing is easily performed, but these splices have a lower reliability due to their mechanical vulnerability.

Fusion splicing needs special fusing equipment, but these splices have a lower loss and a much better reliability than mechanical splices. Automatic fusion equipment exists, offering high-performance splices with an insertion loss of less than 0.1 dB and a reflection below −70 dB. It can be concluded that fusion splices are the favourable type for ODNs.

2.2.1.5 *Optical connectors*

Optical connectors [2] will be used at the interfaces where non-permanent connections are required. Such connections are required to make measurements possible, e.g. optical output power of the terminal equipment. Therefore connectors are generally present between the ODN and the terminal equipment.

The optical connectors used in high bit rate systems have to fulfil the following requirements: low insertion loss, low reflection, high stability, and high repeatability. The only way to fulfil these requirements is by making physical contact between the fibres. There are several types of physical contact connectors. With these connectors reflection values of −45 dB to −60 dB and insertion losses below 0.5 dB are obtained.

2.2.1.6 *Optical network loss*

An indicative value of the total loss for a PON in the 1300 nm window will be given. The PON configuration chosen for this loss calculation

Table 2.2.2
Optical network loss

component	minimum loss	maximum loss
fibre	0 dB	5 dB
splices	0 dB	2 dB
splitters	14 dB	19 dB
WDMs	0 dB	2 dB
connectors	0 dB	1 dB
total network loss	14 dB	29 dB

is frequently used for such estimates. This PON has an ODN with separate fibres for upstream and downstream transmission, a split ratio of 32 (5 couplers with a splitting factor of 2), a length of between 0 km and 10 km, and a maximum of 2 splices/km. Two connectors are used to connect the termination equipment with the ODN. A loss calculation of such a network, based on the figures described in the section earlier, is given in Table 2.2.2.

The loss calculation for such a network will be based on the component losses mentioned in the previous sections, except for the fibre. For the fibre loss a realistic maximal of value 0.5 dB/km will be taken. The minimum loss is calculated for 0 km path length and minimal component losses. The maximum loss is calculated for 10 km with the maximum number of splices and maximal component losses. For the minimum loss case no WDMs will be taken into account, for the maximum loss case 2 WDMs will be present. The assumptions above are taken for the loss calculation to obtain a realistic value of the optical path loss and its variability for the optical network supposed.

As can be seen from the table, the optical loss of such a network lies between 14 dB and 29 dB. Having a certain transmitter power, the receiver has to cope with the whole range of the loss variability. A regular high-power optical transmitter has an output power of 1 mW or 0 dBm. Combining this figure with the optical path loss values as calculated, the receiver has to handle an input level between −14 dBm and −29 dBm. The level variation will occur in bursts; each received signal can have a level lying anywhere in that range. The burst mode consequences for the receiver design will be discussed in Chapter 3.4.

2.2.2 Active optical components

In an optical communication system, the selection and specification of optical components is mainly determined by the transmission parameters of the network and the modulation technique applied. The two most important network transmission parameters are the attenuation and the dispersion of the optical path between the transmitter and receiver.

At bit rates below 1 GHz, using an intensity modulation direct detection (IM-DD) system at 1300 nm over SSM fibre, deteriorating effects such as dispersion and laser noise have hardly any influence. For such systems, the ODN loss can be considered as the requirement for the power budget for transmitter and receiver.

Properties of the active optical components discussed in this subchapter are related to the IM-DD transmission system, because actual PON systems for interactive traffic are all applying this technique. Other modulation techniques may demand special requirements that are not within the scope of this section. This subchapter describes the major properties of active optical components needed to construct PONs, i.e. optical sources, optical detectors and optical amplifiers.

2.2.2.1 *Optical sources*

Possible optical sources to be applied in a communication system are the laser diode and the light-emitting diode (LED). Both of these optical sources have been extensively discussed in the literature; see e.g. [1].

In both diodes light is primarily generated by the recombination of electrons and holes inside a forward biased PN-junction. This mechanism is called spontaneous emission. The bandgap energy of the material determines the wavelength of the generated light. Modern lasers have, however, much more complicated internal structures than only a simple PN-junction. In a laser, the main light generating process is stimulated emission. Photons trigger the generation of additional photons by stimulating additional recombinations. The stimulated photons are coherent (having the same wavelength and phase) with the generating photons. This results in a coherent light output with a relatively narrow spectrum width.

The fundamental difference between the LED and the laser is the optical resonator that is used in the laser. In the Fabry-Perot (FP) laser the resonator is formed by differences in the refractive indices of the different layers, combined with semi-transparent mirrors at the facets that terminate the optical cavity or FP resonator. More recent laser types use more sophisticated resonator constructions thereby improving in particular the spectral behaviour of the emitted light. For IM–DD systems in the access network this is not a critical issue; instead, cost is the prevailing argument and, therefore, FP lasers present the most appealing option.

The output spectrum of an FP laser shows a number of discrete output frequencies (modes) with a total width of some nm. Other characteristics of the FP laser are the fast modulation speed (up to 10 GHz) and the narrow angle of the output beam, which makes coupling power into the fibre rather easy, resulting in an obtainable output power of about 1 mW. Compared with the laser, the LED has a much lower modulation speed, its output spectrum is much wider and the radiant intensity is considerably lower. For these reasons the LED is no alternative to the FP laser in high-speed IM–DD access systems. Important features of packaged laser diodes are the built-in monitor diode and the Peltier temperature control element. With the aid of these components, the output power can be stabilised and made independent of temperature and aging effects.

2.2.2.2 *Optical detectors*

Optical detectors convert the received optical power into an electrical current. This current is amplified and processed to deliver information in a useful format. Suitable types of detectors for PONs are the PIN-diode (p-material, intrinsic layer, n-material) and the APD (avalanche photodiode).

The principle of operation for both diodes is based on the absorption of the energy of a photon that is entering the intrinsic layer [1]. Absorption only occurs if the photon energy is larger than the bandgap energy of the material, causing the generation of an electron–hole pair in the intrinsic layer of the diode. The ideal photodiode generates one electron–hole pair per photon. The relation between the generated electrons, contributing to the photocurrent, and the incident photons is called quantum efficiency, which ideally has a value of 1. The usual way to express the photosensitivity is as responsivity, defined as the photocurrent divided by the incident optical power. This value is proportional to the wavelength because the energy of the photon is inversely proportional to the wavelength. The wavelength where the photon energy is equal to the bandgap energy, is called the cut-off wavelength. Above the cut-off wavelength, an abrupt drop in responsivity occurs. The PIN-diode is usually applied with a reverse bias voltage, in which case a certain leakage or dark current is flowing.

In an APD, the principle operation described above is combined with an internal carrier multiplication. This is obtained by applying a reverse voltage to the diode near the breakdown value. In this way a higher responsivity is obtained than with PIN-diodes. The APD is constructed with an extra semiconductor layer compared with the PIN-diode. At a junction of this layer a large electric field causes the multiplication effect. The multiplication factor, being typically about 10, is strongly dependent on the applied voltage and temperature. Stabilisation is therefore required. Another APD effect is the multiplication of the dark current.

The sensitivity of an optical receiver is determined by the photodiode and the technology of the amplifier connected to the photodiode. For a direct detection optical receiver at e.g. 622 Mb/s the sensitivity range required to obtain a bit error rate of 10^{-9} is about −30 dBm to −40 dBm, depending on the type of photodiode and amplifier technology. This value is valid for regular, continuous level receivers. For burst mode reception, which is the case for upstream traffic in TDMA PONs, the sensitivity of the receiver will deteriorate by some dBs (see Chapter 3.4). With a transmitter power of 0 dBm, the obtainable power budget based on the figures above is 30 dB to 40 dB. Comparing these values with the earlier given maximum network loss value of 29 dB shows that a PON as described is feasible at 622 Mb/s.

2.2.2.3 Optical Amplifiers

Current practices are focused on using optical amplifiers in long-distance trunk systems, where a one-to-one replacement of regenerative electronic repeaters with fibre amplifiers may take place. But the optical amplifier, in all its various realisations, can also offer opportunities in the local loop [1]. In general, an optical amplifier can be used in optical networks as a booster, as an in-line amplifier or as a preamplifier. For the subscriber loop, especially when a tree and branch network is considered, the most attractive option is to utilise the optical amplifier as a booster to compensate for network losses.

Two types of optical amplifiers are discussed: the active fibre amplifier (AFA), and the semiconductor optical amplifier (SOA). The AFA consists of a pump laser, a WDM and a length of active fibre. The active fibre is doped with a special material that allows for amplification if pump power at a certain wavelength is applied. For 1550 nm, Erbium doped amplifiers are reasonable mature and are commercially available now. They can have a gain of up to 25 dB and a maximum output power of about 10 dBm. For 1300 nm, research is still going on, based on Neodymium and Praseodymium doping. These types are not yet available.

The work being done on SOAs is small compared with the work on AFAs. SOAs in the 1550 nm window are, in many ways, inferior to their Erbium doped counterparts; exhibiting lower gain, lower saturation power and polarisation dependency. The reduction in the available gain, due to the high coupling losses and the generation of intermodulation products owing to inherent nonlinearities, has caused many people to switch their attention to fibre amplifiers. The latest developments on SOAs, however, indicate that the polarisation dependency and coupling problems can be largely overcome by paying careful attention to device geometry.

REFERENCES

[1] Mestdagh D J G, *Fundamentals of Multiaccess Optical Fiber Networks*, Artech House: Boston 1995
[2] Kashima N, *Optical Transmission for the Subscriber Loop*, Artech House: Boston 1993

2.3 MULTIPLE ACCESS METHODS

2.3.1 Introduction

A multiple access method is a technique that is used for sharing a limited resource amongst a number of users. In a PON access system, the shared resource is the communication bandwidth. Users vying for this resource may include subscribers, service providers, operators and various network entities. There are different multiple access methods suitable for PONs. Most of these methods have already been used in traditional, electrically based systems for many years. The WDMA technique is used exclusively in optical communications.

Each multiple access method takes advantage of some characteristic of a communication system in order to divide and allocate the available bandwidth. Table 2.3.1 lists the different access methods and the specific resource characteristic that each method utilises. This chapter discusses the techniques listed in Table 2.3.1 and highlights any advantages or disadvantages that may be associated with each technique. Although different multiple access methods can be used in combination, this chapter will only elaborate on each individual technique. Coherent techniques will not be discussed in this chapter because they are still too expensive to be used efficiently as a multiple access method in a PON.

2.3.2 TDMA

Time division multiple access (TDMA) is a technique that divides the available transmission bandwidth into sequential time slots. Time slots may be preassigned to users or they may be allocated on a demand basis, depending on the type of transfer mode being used. Two typical types of transfer mode are STM and ATM.

The synchronous transfer mode (STM) assigns fixed time slot intervals to a user. Each user is therefore guaranteed a specific amount of available bandwidth. Time slots may be statically allocated on a semi-permanent basis by the system operator, or they may be dynamically allocated by the system for the duration of a call. In either case, the user maintains control of a particular time slot that may not be used by another.

Table 2.3.1

Multiple Access Method	Resource Characteristic
TDMA	Time
WDMA	Optical Wavelength
SCMA (FDMA)	Electrical Frequency
CDMA	Code

Access to B-ISDN via Passive Optical Networks. Edited by U. Killat
© 1996 John Wiley & Sons Ltd

The asynchronous transfer mode (ATM), on the other hand, assigns time slots to users based on the actual demand for data transfer. By dynamically assigning time slots, the total bandwidth can more efficiently be shared among multiple users. In comparison with STM, ATM requires more information regarding communication services and traffic behaviour to ensure that the bandwidth is fairly used.

Figure 2.3.1 shows an example of a TDMA system on a tree PON. Each user is allowed to send data upstream in a particular time slot. The demultiplexer sorts the incoming data according to either its time slot position or some information sent in the time slot itself. Downstream transmissions are also sent in specific time slots.

An important characteristic of a tree PON using TDMA is the obligatory synchronisation of upstream traffic to avoid data collisions. These will occur if two or more data packets from different subscribers arrive at the same time at an optical coupler. The signals are superimposed upon one another and create a composite optical signal. The head-end cannot correctly interpret this signal and a series of bit errors and the loss of upstream information result.

To avoid such collisions taking place, the network must ensure that each subscriber terminal sends its information at a time that allows for

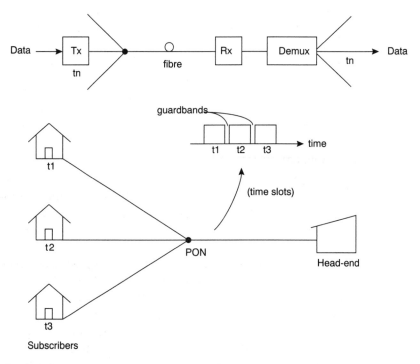

Figure 2.3.1
TDMA PON system

data interleaving. Each subscriber terminal therefore needs knowledge of both a particular time slot and the actual system time when that time slot occurs. The time slot information is provided by the medium access control (MAC) protocol and the system time is provided in the form of delay information that is sent to each terminal from the head-end. The delay information tells each terminal how long it must wait before sending information in a particular time slot. This requires a knowledge of the distance between each subscriber terminal and the head-end. Once the distance is known, the delay time calculation can be made with respect to a downstream time slot.

The process whereby the exact distance between each subscriber terminal and the head-end is measured and recorded is called ranging. Based on this measurement, delay intervals can be sent to each subscriber to ensure that data collisions in the upstream direction do not occur. Spaces inserted between each upstream transmission are called guardbands. Guardbands allow for small variations in the data transmission times which would otherwise cause data collisions.

Synchronisation techniques are also required to locate the beginning of each time slot. This is necessary because the ranging process is not accurate enough to determine the exact starting time of each slot. Given that the data transmission speed is known, the synchronisation procedure at the head-end must determine the starting position, or clock phase, of the data burst in each time slot. Synchronisation can be achieved through the use of, e.g., a phase-lock loop (PLL). Special control signals or preambles in the data stream may be employed to permit the head-end to synchronise the data bursts with its internal clock.

The optical power level is another important aspect in the design of a TDMA tree PON. Upstream data transmissions from different subscribers will arrive at the head-end's receiver with differing optical power levels. This variation is due to the different losses experienced by the optical signals as they pass through different branches in the PON. The head-end's receiver must therefore be able to quickly adjust its logic threshold for each burst so that it can correctly distinguish a "0" from a "1" . This functionality is commonly provided by a burst mode receiver.

TDMA is an established method for sharing bandwidth and is found in practically all transmission networks. A TDMA-based tree PON can use a single optical wavelength for upstream and downstream communication, thereby saving other wavelengths for other services. Although a TDMA PON system is not able to fully exploit the potential bandwidth of the optical fibre, it is a relatively inexpensive system to implement and maintain, thus making it a good choice for access networks.

2.3.3   WDMA

Wavelength division multiple access (WDMA) takes advantage of the different optical outputs that can be generated by semiconductor light sources. These sources can be fabricated to generate light at specific wavelengths. Different wavelengths can be optically combined on the same fibre and then later separated to create a multichannel multiple access system.

The term WDMA traditionally refers to a system that uses optical sources falling into different optical windows. Optical windows are wavelengths in an optical fibre that allow for the transmission of light with little loss or attenuation. There are essentially three optical windows for silica fibres: one at 850 nm, 1300 nm, and 1550 nm. If, however, the different wavelengths are spaced closely together within the same optical window, then the technique is sometimes referred to as HDWDMA, or high-density WDMA. Although these two techniques are different regarding their implementation, there is no conceptual difference between the two, so the term WDMA will be used for both situations. WDMA networks can be broadly divided into two different categories: single-hop (all-optical) networks and multihop networks.

2.3.3.1   *Single-hop WDMA*

Single-hop WDMA networks transmit a data message from the source node to the destination node in one step, without the need for an optical-electrical-optical conversion and retransmission. Single-hop WDMA networks can be divided into two categories: the broadcast-and-select and the wavelength-routing networks.

A broadcast-and-select network operates by distributing the combined optical signals to all of its receivers via an optical coupler. The selection process works by using tunable senders and/or receivers to couple the transmitters with their respective receivers. Three possibilities exist for this selection process.

- Tunable transmitters and fixed receivers. The transmitter is tuned to the desired receiver's frequency. Data collisions may occur if two or more transmitters simultaneously send information to the same receiver.

- Fixed transmitters and tunable receivers. The receiver is tuned to the desired transmitter's frequency. This method has the advantage of permitting multicast messages. Messages may be lost if the receiver can only tune to one transmitter at a time.

- Tunable transmitters and tunable receivers. The transmitter and the receiver can both be tuned to a common frequency. This is the most flexible method for establishing a communication channel, allowing for different combinations of point-to-point, point-to-multipoint, and multipoint-to-point connections.

In all cases, a protocol is required if dynamic connection establishment/release is desired. The protocol must also ensure that data collisions and lost messages do not occur. This may require the use of input/output buffers to delay the transmission or processing of messages. The most flexible single-hop WDMA architecture for connection establishment — i.e., using tunable transmitters and tunable receivers — is also the most costly and complicated method to implement in terms of components and protocols. The wavelength-routing network uses passive, wavelength selective devices to route information from the transmitter to the receiver. A connection is established by combining the correct transmitter frequency with the correct network path to the receiver. The network may use fixed or programmable wavelength devices for routing the optical signal. If, however, fixed devices are used for network routing, tunable transmitters/receivers are necessary to dynamically alter the connections.

The main source of noise for a single-hop WDMA network is optical crosstalk. Optical crosstalk is the interference caused by one optical signal to another. This interference may occur in the fibre itself or in the receiver portion of the network, and is referred to as nonlinear or linear crosstalk, respectively.

Nonlinear crosstalk can be neglected in the context of PONs.

Linear crosstalk, on the other hand, is the result of undesirable characteristics in the channel selection method. Linear crosstalk depends on the type of modulation used (coherent/non-coherent) and the inter-channel spacing. For coherent systems, the crosstalk depends on the optical linewidths of the transmitting and the local lasers, and the electrical response of the bandpass filter. For non-coherent systems, the crosstalk depends on the response of the optical filter that is used. Given that the amount of crosstalk can be estimated for any particular system, the minimum inter-channel spacing can also be calculated.

Linear crosstalk may also occur if the stability of the laser diodes or the optical filters is poor. Instabilities will cause the optical wavelength to vary and possibly interfere with neighbouring channels. Laser diodes are particularly susceptible to variations in temperature and in the injected current, while optical filters are susceptible to temperature changes. Different techniques have been developed to help stabilise the wavelength of an optical signal as it is generated by a laser or passed through a filter. Many lasers, for example, use thermo-electric cooling devices to control their temperature. Wavelength locking mechanisms are used by lasers to prevent variations in the generated optical wavelength. For filters, new materials are being employed that reduce temperature related instabilities.

Combined with other factors, such as the sensitivity of the receivers, crosstalk limits the total number of channels that can be multiplexed in a given system.

2.3.3.2 *Multihop WDMA*

Multihop WDMA networks require one or more intermediate nodes to transmit the optical data from the source node to the destination node. Each node uses unique input and output wavelengths for reception and transmission. Each intermediate node, or hop, makes an optical-electrical-optical conversion of the data. The data is switched through the network — from node to node — until the destination node is reached. This method is similar to that of traditional packet switching networks. The application of unique transmission and reception frequencies for each node removes the need for tunable receivers and transmitters.

The throughput of multihop networks can vary greatly, depending on the network topology and the characteristics of the data traffic. A multihop network can be mathematically modelled as a directed graph, allowing for throughput and delay calculations to be made on different network proposals.

2.3.3.3 *WDMA PON*

Figure 2.3.2 shows a WDMA tree PON system that is based on a single-hop, broadcast-and-select type of network. Each subscriber sends data upstream using a unique optical wavelength. The signals are coupled in the optical coupler and carried to the head-end. The head-end separates the different optical signals using a wavelength division (de)multiplexing (WDM) device. Several different techniques can be used for demultiplexing, e.g. optical filters and gratings. The separated wavelengths are sent to different optical receivers and converted back into electrical pulses. Downstream traffic can also be sent and demultiplexed using this approach.

Although WDMA systems can theoretically take advantage of the large bandwidth available on an optical fibre, they are currently limited by the performance of key optical components. Semiconductor lasers and optical filters, for example, are limited by their operating range, speed, and response over a wide range of wavelengths. The use of complex and expensive optical components does, however, simplify the design of the electrical circuits. Single hop WDMA systems are well suited to circuit switched transmission and routing functions, while the multihop WDMA architecture is best suited to packet switched networks.

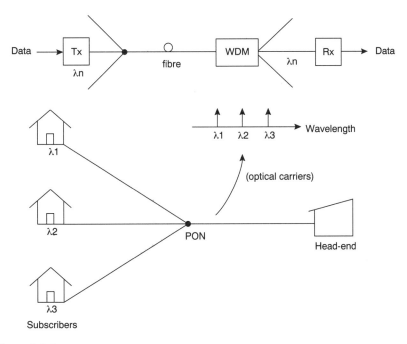

Figure 2.3.2
WDMA

2.3.4 SCMA

Subcarrier multiple access (SCMA) is a technique that uses different electrical subcarrier frequencies to multiplex/demultiplex separate data streams. One or more optical carriers are then intensity modulated by the different subcarriers. SCMA systems can be classified as either single-channel SCMA or multichannel SCMA.

2.3.4.1 *Single-channel SCMA*

In a single-channel SCMA system, each data stream modulates a separate, electrical subcarrier frequency; each subcarrier frequency then modulates a separate optical carrier signal. Unless WDMA is being applied, the different optical carriers will be of the same approximate wavelength. The optical carriers may be combined with an optical coupler to form a composite optical signal. The performance of a single-channel SCMA system is limited by the overall signal-to-noise ratio (SNR). Factors that can affect the sensitivity of the receiver, and thus the overall SNR, include shot noise, optical beat interference, thermal noise, and fluctuations in the laser intensity.

Shot noise is a quantum effect that randomises the optical–electrical conversion of photons according to Poisson statistics. The statistical distribution of the arriving photons means that the exact conversion time of a photon has an associated degree of error (noise). Shot noise is dependent on many factors, including the optical signal power and the type of photodiode being used. Shot noise increases as the number of channels (transmitters) in a system increase.

Optical beat interference (OBI) is an intermodulation product that arises during the conversion of multiple optical carriers into an electrical signal by a photodetector. Each pair of optical carrier signals produces an intermodulation product at a frequency equal to the difference between the optical carriers' centre frequencies. If an intermodulation product happens to fall into the frequency range of a desired subcarrier signal, it will degrade the SNR of that signal. OBI can be avoided by using optical carriers at different frequencies (WDMA), or it can be overcome by increasing the modulation index of each optical carrier.

Thermal noise, also referred to as electronic noise, is the result of the random motion of electrons in a conductor, and occurs in the receiver front-end. This noise type is dependent on the receiver's design and technology, the load resistance, the bandwidth of the receiver, and the absolute temperature.

Laser intensity fluctuations are primarily the result of undesired spontaneous photon emissions that affect the amplitude and phase of the laser's light. The value of the amplitude fluctuation is defined in terms of relative intensity noise (RIN), which is defined as the power spectral density of the normalised output power fluctuation. This value can be measured and theoretically calculated for different laser types. The RIN value is strongly affected by light reflected in the optical fibre and fibre–component interfaces. Optical isolators are necessary in SCMA and coherent applications, which are sensitive to laser intensity noise.

2.3.4.2 *Multichannel SCMA*

A multichannel SCMA system is similar to the single-channel system, as each data stream first modulates a separate subcarrier. The difference is that the subcarriers are combined in the electrical domain before modulating a single optical carrier. This allows more channels to be multiplexed and transmitted than by using a single-channel SCMA system. Single-channel SCMA systems must use relatively wide optical wavelength spacings to prevent OBI, and thus waste potential communication bandwidth. Multichannel SCMA suffers from the same noise sources that are present in single-channel SCMA, as well as from two new sources: subcarrier intermodulation products and clipping.

Subcarrier intermodulation products are generated by a semiconductor laser that has a multicarrier input signal. The intermodulation effect is the result of device nonlinearities in the laser. If the intermodulation products fall within the bandwidth of a subcarrier, then the SNR for that carrier is degraded.

Clipping results if the drive current to the semiconductor laser falls below the lasing threshold current. If this occurs, all lasing activity will cease and the optical power level will fall to zero, thereby clipping the output power signal off at the low end. Clipping effectively limits the degree to which the subcarriers may be modulated, since a large subcarrier modulation index can result in an input current variation that leads to values below the threshold current.

The receivers for the single-channel and the multichannel SCMA systems will operate in an identical manner. Depending on the receiver type, it will first perform an optical–electrical conversion, and then demodulate a specific subcarrier either by using a bandpass filter and subcarrier demodulator, or by first mixing and then filtering the subcarrier.

2.3.4.3   SCMA PON

Figure 2.3.3 illustrates an example of an SCMA tree PON network. The network operates as a single-channel SCMA system. Each subscriber terminal uses a different electrical subcarrier for carrying its baseband

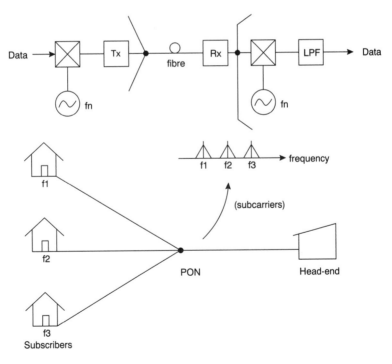

Figure 2.3.3
SCMA PON system; LPF, low pass filter

data. The subcarrier then modulates an optical carrier signal to be sent to the head-end's receiver. The optical carriers from each subscriber are in the same wavelength region and are optically combined in the passive network.

The SCMA receiver consists of a single optical receiver and a number of electrical demodulation circuits, one for each subcarrier transmitter. The difference between this technique and WDMA is that SCMA separates signals based on their differing electrical frequencies, whereas WDMA separates signals based on differing optical frequencies (or wavelengths). The complexity of an SCMA system lies in the electrical components required to multiplex and demultiplex the separate data channels. This removes the need for tunable optical receivers and transmitters, which are much more expensive than their electrical device counterparts. The biggest limitation to using SCMA in an access network is the number of active subscribers that can be served simultaneously. This limitation arises from the previously discussed noise factors, including: shot noise, optical beat interference, thermal noise, fluctuations in the laser intensity, and, in the case of multichannel SCMA, subcarrier intermodulation products and clipping effects.

2.3.5 CDMA

Optical code-division multiple access (O-CDMA) has been studied over the past decade for various applications, but particularly for use in local area networks. Although spread-spectrum techniques are generally known and well understood today [3] there are some problems which are specific to O-CDMA. But before these problems are treated a short overview of direct-sequence CDMA is presented.

2.3.5.1 Code-division multiple access

CDMA is a spread-spectrum multiple access technique that allows many users to transmit information simultaneously (as in FDMA) on the same electromagnetic carrier frequency (as in TDMA) and using the same physical medium. Orthogonality between the different connections is achieved by using an orthogonal code set, assigning one codeword to each connection. The bits of a codeword are called chips. Every payload bit is now encoded by a certain number of chips; this number of chips is called the spread factor. In coherent (radio) CDMA systems bipolar codes can be used. Gold sequences and maximum length sequences are examples from which balanced bipolar codes can be generated [4,5].

User data from a data source are coded by multiplication with a code sequence running a factor n (the spread factor) faster than the user data. This signal is fed to the channel, i.e. the common physical medium, where noise may be added. The noise consists mainly of interference

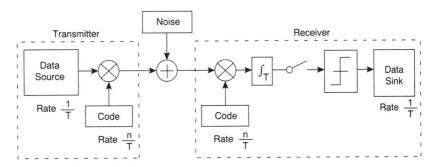

Figure 2.3.4
Principle of direct-sequence CDMA

from other simultaneously active transmitters. The channel signal is then decoded by means of a correlation receiver using the same code as the corresponding transmitter, cf. Figure 2.3.4. Receiver synchronisation, which generally is an important aspect of CDMA systems, will not be treated here since it is not specific to optical CDMA. The focus is on the specific problems of optical CDMA.

2.3.5.2 Optical CDMA

Optical CDMA can be divided in two categories:

- Incoherent O-CDMA
- Coherent O-CDMA.

The assumption that the optical transmission is accomplished with intensity modulation and direct detection (IM/DD) leads to an incoherent transmission system. This has severe consequences for the CDMA code to be chosen. On the other hand, if it is desirable to benefit from the better performance of coherent CDMA, a coherent transmission system, which implies the technological problems of coherent optical communications, must be supplied.

2.3.5.3 Incoherent O-CDMA

Incoherent transmission, as it is used in most of today's optical communication systems, is based on intensity modulation of the optical source and on direct detection of the optical output power of the channel. So the channel is a unipolar channel[†]. This means that true orthogonality between the codewords of a code could only be achieved by using a set of codewords without overlap between "light on" chips with the additional requirement of perfect synchronisation between code generators of the involved transmitters. In this respect TDMA can be regarded as one special synchronous CDMA system with a code

---

[†] Also called "OR channel"

consisting of codewords with a weight of only one chip. Generally, one has to live with the fact that only quasi-orthogonal codes can be found for incoherent CDMA. Such codes are called optical orthogonal codes (OOCs) [6]. A $(n, w, \lambda_a, \lambda_c)$-OOC is a set of (0,1)-sequences of length n with weight w that meets the following correlation constraints.

- **Autocorrelation property** The autocorrelation of any codeword is at most λ_a for any time offset different from zero. This guarantees that a codeword can be distinguished from a time shifted copy of itself.

- **Cross-correlation property** The cross-correlation between any two codewords is at most λ_c for any time offset. This guarantees that codewords can be distinguished from each other.

The practical usefulness of OOCs is severely limited by some contradictory requirements: on the one hand the code weight must not be too small in order to enable detection of the autocorrelation peak within the noise generated by the multiple access interference (MAI), on the other hand the weight must not be too high — otherwise the code set size is not sufficient. The order of this contradiction may well be seen taking a look at an upper bound to the code set size [6] of $(n, w, 1, 1)$-OOCs for some practically reasonable values of n and w; the theoretical maximum code set size for weights from 3 to 8, and for code lengths from 50 to 2000 chips is shown in Table 2.3.2.

To get large code sets (and thus be able to have many simultaneously active connections in a network) one has to go in the direction of the upper right corner of the table, i.e. high spread factor and low weight. A straightforward optimisation leads to a code with weight 1 resulting in n different codewords. This would require synchronisation of all transmitters, and the result is a TDMA system. If the synchronisation efforts are too high, or if an asynchronous system is desired, a low code weight makes it difficult to decide under heavy multiple access interference, whereas a higher weight is good against the MAI, but does not allow large code sets. Some recent works use more sophisticated coding schemes, e.g. composite codes formed from Barker codes and Gold sequences [7]. This approach delivers better error behaviour, but it requires a relatively complicated coding.

Finally, it can be stated that incoherent optical CDMA has proved to be inferior regarding throughput compared with TDMA [8].

Table 2.3.2
Theoretical maximum size of $(n, w, 1, 1)$-OOCs

$n =$	50	100	250	500	1000	2000
w						
3	8	16	41	83	166	333
5	2	4	12	24	49	99
8	0	1	4	8	17	35

Nevertheless, some recent approaches employing multi-user detection in optical CDMA systems indicate that the relatively poor error behaviour of optical CDMA under high loads can be drastically improved. The benefit of CDMA clearly is in its inherent switching capability which does not require a complicated MAC protocol. CDMA might thus be used in an integrated services network for a relatively high number of low bit rate connections, while TDMA could be used for relatively few high rate connections [9].

The code generation as well as the decoding can be done optically, e.g. with a tapped delay line set-up [10,11]. Here for each user data bit "0" nothing is sent; for each "1" bit an optical pulse with a duration of one chip is fed to the optical coder shown in Figure 2.3.5. This pulse is split by an optical splitter/combiner (S/C) and led through w fibre delay lines with the respective delays for the desired codeword to be used for this connection. The output of the delay lines is then combined to the output fibre and thus forms the CDMA coded signal. The decoding correlator works similarly. Such optical signal processing devices have the advantage of high processing speed and good reliability, and they do not require any additional power supply. There are, nevertheless, some severe disadvantages for the considered CDMA applications: the code is "hard wired" and cannot easily be changed. Moreover, the tapped delay line set-up is rather unwieldy and only practically feasible for low weight codes, because the number of delay lines equals the weight of the code.

However, there are also other proposals for optical encoding and decoding of CDMA, also for coherent CDMA. Optical signal processing is still a field of strong activity, and integrated optics with electronic control should show some progress that might also be useful for optical CDMA. Recently some very promising proposals have been published that suggest technological solutions for interference-free incoherent optical CDMA systems [13]. These systems can achieve a capacity like that of coherent CDMA systems, but without the requirements of coherent optical technology. Thus, they could become one possible multiple access solution for future PONs with large numbers of

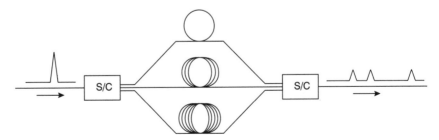

Figure 2.3.5
Passive optical CDMA coder; S/C, splitter/combiner

subscribers where time-division methods will run into problems owing to their lack of flexibility.

2.3.5.4 Coherent O-CDMA

There have been some experiments on coherent optical CDMA [12]. Coherent CDMA does not exhibit the above-explained problems of incoherent CDMA in finding suitable codes. The problems of coherent O-CDMA are on the technological side: expensive optical devices are necessary. This will not be discussed in depth here, because the coding theory for coherent CDMA is a subject already widely covered in the literature, and the high technological and monetary efforts necessary for coherent optical systems make it quite improbable that they will be used in the near future. Nevertheless, coherent optical CDMA could gain some importance with rising bandwidth requirements in future communications demands.

REFERENCES

[1] Mestdagh D J G, *Fundamentals of Multiaccess Optical Fibre Networks*, Artech House: Norwood, MA, 1995

[2] Kashima N, *Optical Transmission for the Subscriber Loop*, Artech House, Norwood, MA, 1993

[3] Pickholtz R L et al, *Theory of Spread-Spectrum Communications — A tutorial*, IEEE Trans Comm, **30**(5), 855 — 84, 1982

[4] Sarwate D V and Pursley M, *Crosscorrelation Properties of Pseudorandom and Related Sequences*, Proc IEEE, **68**(5), 593-619, 1980

[5] Gold R, *Optimal Binary Sequences for Spread Spectrum Multiplexing*, IEEE Trans Inform Theory, **13**(4), 619-21, 1967

[6] Chung F R K et al, *Optical Orthogonal Codes: Design, Analysis and Applications*, IEEE Trans Inform Theory, **35**(3), 595-604, 1989

[7] Zaccarin D and Kavehrad M, *New Architecture for Incoherent Optical CDMA to achieve Bipolar Capacity*, Electron Lett, **30**(3), 258-9, 1994

[8] Walle H and Killat U, *Capacity Comparison of Incoherent Optical CDMA using Prime Codes and TDMA*, ITG-Fachbericht Vol. 130 Codierung für Quelle, Kanal und Übertragung, 393-9, ITG, VDE Verlag: Berlin, 1994

[9] Kwong W C et al, *Performance Comparison of Asynchronous and Synchronous Code-Division Multiple-Access Techniques for Fiber-Optic Local Area Networks*, IEEE Trans Comm **39**(11), 1625-34, November 1991

[10] Prucnal P R et al, *Spread Spectrum Fiber-Optic Local Area Network using Optical Processing*, J Lightwave Technol, **4**(5), 547-54, 1986

[11] Santoro M A, *An Integrated-Services Digital-Access Fiber-Optic Broadband Local Area Network with Optical Processing*, in G Prati (ed) Proc. 4th Tirrenia Int. Workshop on Digital Communications, Tirrenia, Pisa, Meeting on Coherent Optical Communications and Photonic Switching, 285-96, Sept 1989. Elsevier: Brussels

[12] Hajela D J and Salehi J A, *Limits to the Encoding and Bounds on the Performance of Coherent Ultrashort Light Pulse Code-Division Multiple-Access Systems*, IEEE Trans Comm, **40**(2):325-36, Feb. 1992

[13] Zaccarin D and Kavehrad M, *"An Optical CDMA System based on Spectral Encoding of LED*, IEEE Photonics Technol Lett, **4**(4), 479-82, 1993

2.4 MULTIPLEXING FOR BIDIRECTIONAL COMMUNICATION

2.4.1 Introduction

Interactive communication requires bidirectional transmission. There are several methods by which bidirectional transmission on optical fibres can be realised. The following chapter lists several techniques using one or two fibres for point-to-point connections. However the same techniques can be applied to tree structured PON systems as already indicated in the previous chapter describing multiple access techniques. A combination of the given techniques is also possible, but this chapter only discusses the individual systems. The final choice depends on the required upstream and downstream bit rate, network structure, and for PON systems also on the multiple access method.

2.4.2 Two-fibre systems

The simplest solution for bidirectional transmission is the use of the space division multiplexing (SDM) technique. This is realised by the use of separate fibres for upstream and downstream transmission, see Figure 2.4.1. The physical separation of the transmission directions avoids the influence of optical reflections in the network and also avoids the problem of combining and separating upstream and downstream transmission. This in its turn allows for a larger power budget for the optical network, which is of prime importance in the case of a PON tree. By using two fibres more flexibility in network design is made available, because it is possible to extend the network by using WDMs in one or both fibres. This allows for a gradual evolution of new services in the future. By using the same wavelength for both transmission directions the same transmitters and receivers can be used; this reduces the costs for the opto-electronic components.

The main disadvantage is that the number of fibres, splices and connectors used is doubled and in the case of a tree-type PON also the number of optical couplers in the network is doubled. However the costs of fibre cable, passive components and splicing techniques are still decreasing and will probably have a minor impact on total system cost.

2.4.3 One-fibre systems

By combining upstream and downstream transmission only one fibre for each connection to the customer is needed, see Figure 2.4.1. Therefore the amount of fibre, splices and connectors and in a tree PON also the number of couplers in the network is minimised. However

Access to B-ISDN via Passive Optical Networks. Edited by U. Killat
© 1996 John Wiley & Sons Ltd

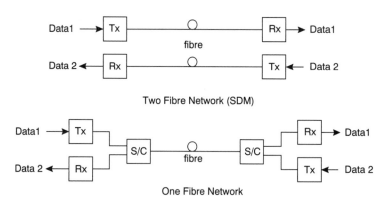

Figure 2.4.1
Multiplexing for bidirectional communications, S/C, splitter/combiner

precautions must be taken against reflections in the network and the available power budget is reduced because of the losses in the optical combiners/splitters. To avoid reflections in the network more advanced and so costly optical connectors, couplers and splicing techniques must be used. It is also possible by using special multiplexing techniques to reduce the influence of the reflections. Common practice is a combination of reducing the reflections in the network and using special multiplexing techniques.

The one-fibre systems can be divided in systems using the same wavelength (DDM, TCM, CDM, SCM) and systems using different wavelengths (WDM) for upstream and downstream transmission. The Table 2.4.1 (subsection 2.4.3.6) lists the different multiplexing methods and the resource characteristics that each method uses.

2.4.3.1   Directional division multiplexing (DDM)

In DDM systems upstream and downstream transmission use the same wavelength. The combining of both transmission directions is done by the use of optical couplers. Therefore the available power budget is decreased by some 7 dB by using couplers on both sides of the connection with respect to SDM. A DDM system is very susceptible to interference caused by optical reflections. Optical reflections are a consequence of imperfections of the passive optical components in the network. Two kinds of reflections are distinguished: back scattering caused by the fibre itself and reflections due to discrete reflection points, e.g. couplers, connectors and splices. For instance, if a power penalty of <0.2 dB is allowed, the ratio of wanted power versus reflected power must be >15 dB for a 622 Mb/s system. The discrete reflection points are the main problem in a DDM system, especially the first optical connector after the coupler, which causes the so-called near-end crosstalk. For example, the ratio of 15 dB, mentioned above for a 622 Mbit/s system, of wanted power versus reflected power requires that the reflection of the first connector must be ≥ 35 dB assuming the transmitters at the OLT and ONU have the same optical output power and the network loss is 20 dB. In this example the other reflections in the network and the passive and active optical components in the terminating equipment are neglected. The reflection problem of the discrete reflections can be reduced by careful selection of the components; thus however will lead to higher system costs. Also the network must be maintained very carefully to keep reflections low which makes DDM unattractive to implement.

2.4.3.2   Time compression multiplexing (TCM)

To prevent the system against degradation by the reflection of the transmitted signals, time compression multiplexing techniques can effectively be used. In TCM systems, upstream and downstream data is transmitted at the same wavelengths in bursts. These bursts are transmitted after each other (so-called ping-pong transmission), thereby preventing reflections of the transmitted signal deteriorating the received signal. During one TCM frame first the OLT transmits a burst while the ONU listens, then the ONU transmits a burst while the OLT listens. In both the OLT and ONU, memory is required to convert the bursty data back into a continuous data stream. In between the bursts there is a silent interval, whose length is at least the round trip propagation delay of the optical signals in the fibre. The required line bit rate is more than twice the bit rate of the source bit rate. For

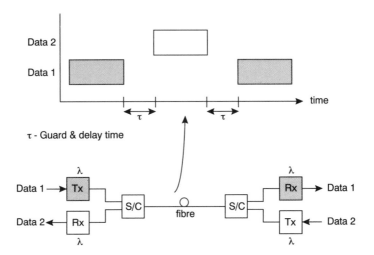

Figure 2.4.2
Time compression multiplexing; S/C, splitter/combiner

instance in a system with a TCM frame of 1 ms, a length of 10 km and a guard time of 5 μs the required line bit rate is 2.25 times the bit rate of the source. Because the transmission is not continuous each burst needs a preamble that contains bits for adjusting the optical receiver and some framing information. This implies that the bit rate in the burst is increased even more which makes the system only suitable for low bit rate applications up to some tens of Mbit/s. As in DDM systems couplers are used to combine both transmission directions which reduces the optical power budget by about 7 dB.

2.4.3.3   Code division multiplexing (CDM)

In CDM systems the effects of reflections of the transmitted signal are reduced by using different codes for modulating upstream and downstream data. Because different codes have little or no correlation the upstream and downstream signals can be distinguished. Therefore the same wavelengths can be used for both transmission directions. However the use of modulation techniques results in a higher line bit rate, which makes it difficult to implement for high bit rates (more than a few 100 Mbit/s) and therefore more costly. As in DDM systems couplers are used to combine both transmission directions which reduces the optical power budget by about 7 dB.

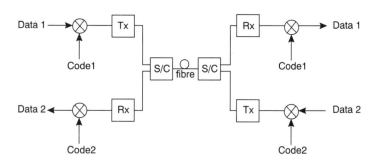

Figure 2.4.3
Code division multiplexing; S/C, splitter/combiner

2.4.3.4   Subcarrier multiplexing (SCM)

In SCM systems the separation of the upstream and downstream transmission is performed by modulating the data with two different (electrical) subcarriers. The resulting signals are modulating the lasers at OLT and ONU which have the same wavelength. At the receiver side after the optical-to-electrical conversion the signal is demodulated by multiplying the signal by an electrical local oscillator. The linearity of the receivers determines the mutual influence of upstream and downstream signals in combination with the reflections in the network.

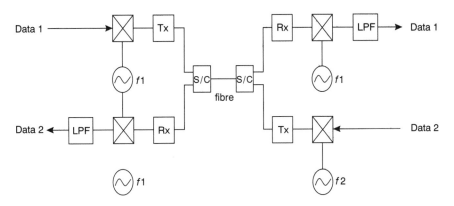

Figure 2.4.4
Subcarrier multiplexing; S/C, splitter/combiner; LPF, low pass filter

Upgrading of the system for other or future services can be done by using another wavelength or another subcarrier. In the case of additional subcarriers the optical power is shared amongst these subcarriers; this reduces the optical power budget. In this case stringent demands are also put on the linearity of the lasers. Techniques and components for SCM are well known and together with the flexibility for upgrading make SCM an attractive system.

The main disadvantages of SCM are the requirements posed on the linearity of the components and in the case of a greater number of subcarriers the reduction of the available power budget. Because of the limited bandwidth for a subcarrier, broadband, e.g. 622 Mbit/s, cannot be accomplished by using a single subcarrier.

2.4.3.5 *Wavelength division multiplexing (WDM)*

In WDM systems multiplexing and demultiplexing is done by using different wavelengths which are combined and separated using WDM devices. Because the attenuation of WDMs is in the order of 1 dB the power budget is reduced by 2 dB compared with an SDM system. The wavelengths used can be in different windows (1300 nm and 1550 nm) or in the same window (e.g. 1300− and 1300+) leaving

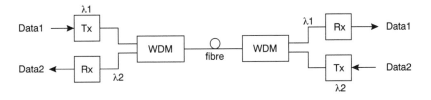

Figure 2.4.5
Wavelength division multiplexing

the other window for other services. More advanced devices are also available which allow for closely spaced wavelengths (high-density WDM-HDWDM). Upgrading in that case can be done by using another wavelength in the same region. The main drawback of a WDM bidirectional system is its sensitivity to reflections in the network in combination with a non-ideal isolation of the WDM. For this reason more advanced WDM devices are required and this increases the costs of the system.

Table 2.4.1
Overview of bidirectional communication systems; A, attention

System	2-fibre	1-fibre				
		DDM	TCM	CDM	SCM	WDM
Multiplexing method	–	DDM	TCM	CDM	SCM	WDM
Resource characteristic	–	–	Time	Code	Electrical frequency	Optical frequency
No of fibres	2	1	1	1	1	1
Combining	–	Coupler	Coupler	Coupler	Coupler	WDM
Network loss	A	A−7dB	A−7dB	A−7dB	A−7dB	A−2dB
Reflection sensitivity	Low	Very high	Medium	Medium	Medium	High
Upgrading	WDM (2)	WDM	WDM	WDM	WDM, SCM	(WDM)

2.4.3.6   Overview of bidirectional systems

The above table gives an overview of the possible bidirectional system as discussed in this chapter. Explanations of the remarks and numbers given in the table can be found in the corresponding subchapters.

REFERENCES

[1]   Green P E, *Fiber Optic Networks*, Prentice Hall: Englewood Cliffs, 1993
[2]   Kashima N, *Optical Transmission for the Subscriber Loop*, Artech House: Boston, 1993

2.5 PONS IN SUPPORT OF ISDN

2.5.1 Introduction

Since the 1980s fibre communication systems have been installed for use in long-distance connections, mainly between trunk exchanges. This application benefited greatly from the peculiarities of fibre transmission: easy transmission of digital binary signals, high capacity, long distances covered without regenerators, immunity to interference. This has led nowadays to a trunk network where fibre has replaced coaxial cables to a large extent.

The situation is very different for the local loop, where distances are very short, signals are of analogue type and a low transmission capacity is required; these circumstances do not seem to justify the introduction of fibre in the access network, as far as residential and small business customers are concerned. In contrast, fibre transmission has already been applied for connection of large business customers, because of their high bandwidth and high security requirements.

Nevertheless there exist at least two main reasons that encourage the introduction of fibre in the access network: the introduction of new services and the integration of different services on the same network. New services requiring higher bandwidths, compared with the plain old telephone service (POTS), are in fact emerging, such as video on demand, video retrieval, videotelephone, video games. They rely on the possibility of delivering digital video and data at high bit rates, of the order of some megabits per second, that are not supported by the existing copper-based access network. Furthermore different networks exist nowadays for each type of service; public switched telephone network, video broadcast network, different types of data networks at various bit rates and using different protocols). Integration of all these types of service into the same network would simplify network management and maintenance and allow flexibility both in the introduction of new services and in the delivery of already existing services to the customers.

Optical fibre transmission can be a valid solution for achieving the targets just mentioned. However, optical transmission technologies developed so far for use in trunk networks have to be optimised for transmission on the local loop to achieve cost-effectiveness. Many solutions have been proposed in the literature, almost always based on the concept that sharing optical transmission equipment among customers reduces the cost per customer connection, making fibre an attractive solution for transmission in the local loop.

PON systems are therefore considered as one of the most promising solutions for the introduction of fibre in the access network. They have been experimented with in the last few years in several laboratory

Access to B-ISDN via Passive Optical Networks. Edited by U. Killat
© 1996 John Wiley & Sons Ltd

and field trials around the world and have already been installed by some network operators. The rest of this chapter is devoted to the presentation of some of these trials, with a particular focus on those that can be considered as milestones in the development of the PON technology for the local loop.

2.5.2   British Telecom TPON system

Around 1988 a PON system for the delivery of telephone and ISDN services (usually referred to as TPON system) was conceived and subsequently developed as a laboratory demonstrator at British Telecom Research Laboratories [1,2,3,4]. The system is based on a tree PON topology with one fibre for both directions of transmission and uses the TDM/TDMA method (described in Chapters 2.3 and 2.4). Fibre-to-the-home (FTTH), fibre-to-the-building (FTTB), and fibre-to-the-curb (FTTC) configurations are supported. The system block diagram is shown in Figure 2.5.1. Voice channels coming from the exchange are part of a 2 Mbit/s PCM stream. An 8 kbit/s signalling channel is derived from the signalling information, contained in time slot 16, for each of the thirty 64 kbit/s voice channels of the PCM frame, thus converting the common-channel signalling to a channel-associated signalling. All voice and signalling channels are then multiplexed

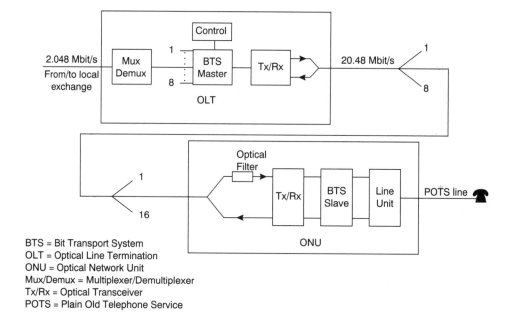

BTS = Bit Transport System
OLT = Optical Line Termination
ONU = Optical Network Unit
Mux/Demux = Multiplexer/Demultiplexer
Tx/Rx = Optical Transceiver
POTS = Plain Old Telephone Service

Figure 2.5.1
TPON system block diagram (adapted from [1])

together to form a single 2.16 Mbit/s stream that, with the addition of some spare bits, gives a 2.352 Mbit/s stream. Eight streams are thus generated and fed to the core part of the system, the so-called BTS (bit transport system) master, where they are multiplexed bit by bit into a single stream. The BTS master adds proper control channels to control transmission over the PON and generates the stream to be transmitted onto the optical network at a bit rate of 20.48 Mbit/s.

Following [6] the units providing the network–side interface and the user–side interface of the optical access network are referred to as optical line termination (OLT) and optical network unit (ONU), respectively.

The ONU filters out only those channels of the TDM frame which are allocated to it. The granularity of the transport system is 8 kbit/s, i.e. ONUs can have access to channel capacities which are a multiple of the basic 8 kbit/s channel. A telephone channel is thus delivered as nine 8 kbit/s contiguous channels, eight of them for the voice channel and one for signalling. A 2B+D ISDN access requires instead nineteen 8 kbit/s basic channels. A further basic channel is allocated to each ONU for housekeeping messages. The system supports a splitting factor of 128 and has a total capacity of 240 telephone channels or 120 ISDN channels.

Upstream transmission from the ONUs to the exchange is based on TDMA. Multiplexing is performed at the bit level and all ONUs are synchronised to the TDMA frame through a proper two-phase ranging procedure. In the first phase, accomplished only at installation of a new ONU, the distance between the OLT and the ONU is measured on command of the OLT using a pulse signal sent by the ONU during an idle period of the upstream frame (the ranging accuracy achieved after this step is ±1 bit). The second phase of the ranging procedure is carried out continuously in order to keep all ONUs always synchronised with the TDMA frame. ONUs transmit the ranging pulses in time-slots reserved in the TDMA frame for this purpose. The OLT monitors the arrival time of the ranging pulses and sets the proper ranging delays in the ONUs to achieve an accuracy of one tenth of a bit. Also, the OLT monitors the received signal amplitudes and adjusts consequently the laser power of the ONUs, in such a way that the signal level at the OLT receiver is equalised to a fraction of a decibel.

In the upstream direction, the channel allocation in the TDMA frame corresponds to the allocation used in the downstream direction in the TDM frame, with the exception of the initial part of the frame that is used for the ONU synchronisation downstream and for ranging upstream. The BTS slave in the ONU allocates the upstream traffic and housekeeping channels in the assigned positions of the TDMA frame. The traffic channels are then extracted by the BTS master in the OLT and demultiplexed into eight separate flows at 2.352 Mbit/s, equivalent to the flows generated in the downstream direction. Finally, the eight 2 Mbit/s PCM streams are formed, by converting the channel-associated signalling into a common-channel signalling, and delivered to the local exchange.

The line bit rate has been carefully chosen as a compromise between the advantages of using a high line bit rate (higher system capacity in terms of line circuits and/or number of customers supported) and the technological limits imposed by: the use of CMOS circuits (chosen for power dissipation constraints); the use of high splitting ratios (for power budget constraints); the difficulty of synchronising many network units with the TDMA frame.

Bit interleaving has been chosen primarily to reduce the complexity of the multiplexing function in the BTS and to remove the need for a Peltier cooler for the ONU laser. With bit interleaving, in fact, lasers emit pulses with a low mark–space ratio.

An upgrade of the TPON system (called broadband PON or BPON) for the delivery of the CATV service is possible through the use of a different wavelength on the optical network. Optical filters are provided for this purpose in the ONUs.

The TPON and BPON systems, together with other optical access systems, were tested in the famous Bishop's Stortford field trial, which covered a two-year period starting in 1990. The field trial mainly

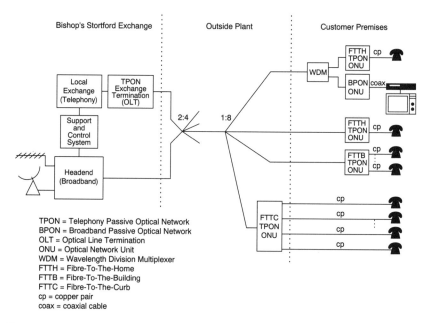

Figure 2.5.2
TPON/BPON field trial architecture

aimed at demonstrating the technical feasibility of the introduction of optical fibre in the local loop and at allowing a comparison of different technical solutions for the optical access network. The architecture of the experimented TPON and BPON systems is shown in Figure 2.5.2. All possible configurations, i.e. FTTH, FTTB and FTTC, were supported in selected areas of the town. Around 200 customers were involved in the trial. Services available were telephony, broadcast TV (18 channels), videotext, video library and digital audio. A service and network management system was included for the management of the field trial.

2.5.3 Deutsche Bundespost OPAL projects

In 1988 Deutsche Bundespost, the German public network operator, began studying new solutions for the introduction of fibre in the access network. This resulted in the series of OPAL (Optische Anschlussleitung–optical access line) projects, with the common objective of testing the identified solutions in several field trials throughout Germany. The following services have been provided to the customers:

- standard POTS
- basic rate (144 kbit/s) ISDN access
- primary rate (2,048 kbit/s) ISDN access
- CATV for 38 or more TV channels, based on the spectrum allocation defined in the BK450 or BK860 standards.

The equipment was provided by different manufacturers for the different field trials. Both country and town areas were involved to include different kinds of user. Table 2.5.1 contains an overview of some trial details.

Optical fibres from OLT to ONUs cover distances ranging from a few hundred metres to some kilometres. The majority of the connections were made according to the FTTC concept, although some FTTB and FTTH connections were also tested depending on the particular field trial situation.

The optical networks are based on the tree topology (except for OPAL 1 which uses a bus topology) and are of passive type. Interactive services are supported through a TDM/TDMA transmission system with gross line bit rate around 30 Mbit/s. In the OPAL4 and OPAL5 trials, bidirectional communication is obtained using WDM on a single fibre for interactive services: the 1.3 µm window is used for transmission from the ONUs to the OLT, whereas the 1.55 µm window is used for transmission in the opposite direction. In the other trials two distinct fibres are used for bidirectional communication in the 1.3 µm window.

Table 2.5.1
OPAL projects main characteristics

Project	City	Manufacturer	Start of service	Services	Characteristics
OPAL 1	Köln	Raynet	May 90	POTS BK450	Bus topology Graded index fibre
OPAL 2	Frankfurt	Raynet	May 92	POTS ISDN 2 Mbit/s	Business customers
OPAL 3	Lippetal	Raynet	Apr 92	BK450	Country areas–low customer density
OPAL 4	Leipzig	Siemens	Nov 91	POTS BK450	WDM for interactive services
OPAL 5	Stuttgart	SEL	Sep 92	POTS ISDN BK450 BK860	WDM for interactive and distributive services Optical amplifiers
OPAL 6	Nürnberg	FAST consortium	Sep 91	POTS BK450	Based on British Telecom TPON concept Purposely developed optical cable and connectors
OPAL 7	Hagen	Bosch	Dec 91	BK450 BK860	Country areas

Distributive services (CATV) are transmitted on a different fibre using the 1.3 μm window and direct analogue modulation of the laser. In the OPAL5 trial, the 1.55 μm window is used in conjunction with optical amplification based on Erbium doped fibre amplifiers (EDFAs).

In the following, some of the OPAL projects are briefly described, with a focus on those which are more oriented to the ISDN interactive services. The reader can find a more exhaustive description in [5].

2.5.3.1 OPAL 1

The first OPAL project, OPAL 1, was aimed at serving residential users with telephone and CATV services. The architecture is based on the FTTC concept, i.e. several customers are connected to the same optical network unit using copper drops. The field trial architecture is depicted in Figure 2.5.3.

The optical network, realised using multimode fibres for interactive services and monomode fibres for CATV, is based on a bus topology, with a bus length of 4 km. A maximum of 24 ONUs are supported by a single optical bus system for a total of 192 telephone customers (8 customers per ONU). The line bit rate is around 20 Mbit/s in the downstream direction and about 40 Mbit/s in the opposite direction, where guard times are inserted between packets transmitted by different ONUs to prevent collisions on the optical network.

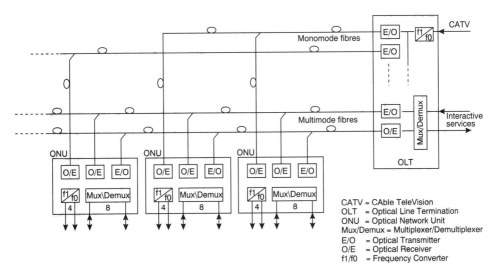

Figure 2.5.3
OPAL 1 field trial architecture (adapted from [5])

Owing to the limited power budget available for the CATV service, imposed by the delivery of the video signals in amplitude modulation format, a smaller splitting ratio is allowed for this service and only three ONUs are fed by a single optical transmitter. Eight transmitters are provided at the exchange site to feed all 24 ONUs. Each ONU supports 4 CATV customer connections, for a total of 96 CATV customers supported by a single system.

This first trial highlighted some problems, such as e.g. the high power consumption of the ONUs, with the need for complicated and expensive remote powering systems, or the necessity of converting the BK450 CATV signal to a higher frequency band to reduce the intermodulation distortion generated by the laser nonlinearities. These problems were solved in the following OPAL projects.

2.5.3.2   OPAL 2

The OPAL2 project was devoted to providing business customers in the Frankfurt banking area with telephone and ISDN connections. No broadband services are supported but, compared with the first OPAL project, OPAL2 has some enhancements as far as interactive services are concerned. First of all, the bus topology has been replaced by a passive double star, using 1:16 optical splitters and the optical network is realised using single-mode fibres. The ONUs are placed at the customer premises and connected to the local power supply, thus avoiding the need for remote powering. Back-up batteries are provided assuring 10–60 min operation in the absence of mains power.

The transmission method is similar to that adopted for the OPAL1 system. The system capacity is 384 64 kbit/s channels, equivalent to a total capacity of 24.5 Mbit/s. Supported services are ISDN primary rate access (2 Mbit/s), ISDN basic rate access (144 kbit/s) and POTS.

2.5.3.3   OPAL 4

In the OPAL4 pilot project a PON is used in the FTTC configuration for providing both private and business customers with broadband distributive and narrowband interactive services. The network architecture is shown in Figure 2.5.4. The CATV signal, after electro-optical conversion, is transmitted by the OLT on the optical network. A 1:2 splitting is used to feed two different ONUs using the same OLT optical transmitter. The distributive signal is then fed to the existing coaxial cable network for distribution towards the customer.

A TDM/TDMA PON system, using a 1:12 splitting ratio, WDM for bidirectional transmission on the same fibre and a line bit rate of 35 Mbit/s, is adopted to support interactive services. The system allows the transport of 2 Mbit/s PCM streams from the OLT to the FTTC ONU. A maximum number of 24 telephone channels is supported per PCM stream, for a total of 288 POTS lines supported by a single OLT. For connection to the customer from the ONU, the existing twisted pair is used. Although only POTS lines are used in the pilot, the system is also able to support ISDN connections.

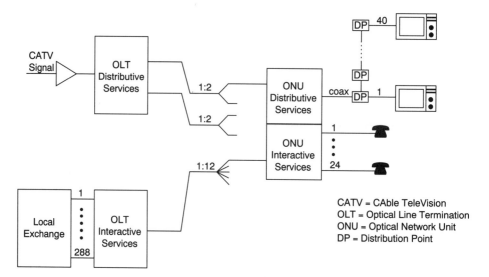

Figure 2.5.4
OPAL 4 field trial architecture (adapted from [5])

2.5.4   Conclusions

Both the British Telecom TPON and BPON field trials and the OPAL
pilot projects have considerably increased knowledge of PON systems.
The most significant issues can be summarised as follows.

- Economical solutions have been found for the introduction of fibre
  in the access network.

- Integration of services is not necessary in a first stage, when technology is not yet mature: it is more convenient to keep distributive
  CATV services separate from interactive POTS and ISDN services.

- Fibre cost is negligible compared with the cost of deploying the
  optical network and the cost of the optical terminations.

- The major problem in analogue CATV distribution is the optical
  power budget on PONs. A high splitting ratio is not achievable
  if sufficient signal quality has to be assured. For this reason, only
  FTTC configurations can be accepted for economic reasons.

Several other field trials required to support POTS, ISDN and CATV
services based on PON technology have taken place in the last few
years around the world. The great interest of the European network
operators in TPON systems has also led to the definition of an ETSI
technical standard [6]. The driving force for introducing this type of
system is the introduction of new services. The existing copper network
will in the near future no longer be able to support more demanding
services using low-cost solutions. In this scenario, fibre seems to open
new horizons, with its almost unlimited capacity.

Plans to introduce PON systems massively are now known from
several network operators around the world. It can be stated that
the first phase of the technological research on PON systems is now
concluded. The second phase has already started and the new challenge is the introduction of the B-ISDN concepts. In fact, research in
the field of broadband PON systems is now oriented towards the introduction of the ATM technique and to the development of systems
with transmission capacities of the order of hundreds of megabits per
second or even gigabits per second. Prototypes of such systems and
even commercial products are already available [7,8], but research is
continuing to increase their performance in the sense expressed before.
The reader can find detailed information concerning the prototype of
an ATM-PON developed in the framework of the European RACE 2024
"Broadband Access Facilities" project in Section 3 of this book.

REFERENCES

[1] Hoppitt C E and Clarke D, *The provision of telephony over passive optical networks*, Br Telecom Technol J, **7**(2), 1989

[2] Stern J R and Hoppitt C E *et al*, *TPON–a passive optical network for telephony*, Proc ECOC 88, 203–6, 1988

[3] Oakley K A and Taylor C G *et al*, *Passive fibre local loop for telephony with broadband upgrade*, Proc ISSLS 88, Section 9.4, 1988

[4] Faulkner D W and Payne D B *et al*, *Optical networks for local loop applications*, J Lightwave Technol, **7**(11), 1989

[5] Sporleder F and Garlichs G *et al*, *Optische Übertragungstechnik für flächendeckende Teilnehmeranschlüsse* Teil 3, Der Fernmelde Ingenieur, October 1992

[6] ETSI prETS 300 463 (DE/TM-3019) *Transmission and Multiplexing; Requirements of passive Optical Access Networks (OANs) to provide services up to 2Mbit/s bearer capacity*

[7] Asatani K, *Lightwave subscriber loop systems toward broadband ISDN*, J Lightwave Technol, **7**(11), 1989

[8] Verbiest W and Van der Plas G *et al*, *FITL and B-ISDN: A marriage with a future*, IEEE Comm Mag, June 1993

Section 3

ACCESS TO B-ISDN VIA PONS

3.1 INTRODUCTION: NETWORK AND SERVICE EVOLUTION

3.1.1 The role of the access network in the evolving telecommunications network

One can think of the global telecommunications network as a network that consists of three main parts: the core network, the customer premises network and the access network. The core network is the part of the telecommunications network where important operations such as switching, billing and transmission take place. During the last few decades, a number of important trends were started that caused a transformation of the core network, including: a shift from analogue to digital switching and transmission; a year-by-year increase in the capacity of switching systems by using more powerful integrated circuits, microprocessors and software; an increase in the capacity of transmission systems by using optical fibre technologies.

One consequence of increasing the capacity of switches is that it creates the possibility of decreasing the number of switches in the field. Another consequence of all these trends is the higher level of integration between services and networks. An evolution is taking place, starting from completely separate networks for e.g. telephony and telex, to data communication via telephone lines (initially by using modems), and finally towards an integrated network for all voice and data services, enabled by the concept of integrated services digital network (ISDN). With other network concepts, such as the (advanced) intelligent network (IN) and the telecommunications management network (TMN), a faster introduction of new services and superior operation and management procedures can be realised by the network operators. Furthermore, halfway through the nineties, the core network is faced with the introduction of new transmission and switching techniques such as synchronous digital hierarchy (SDH) and asynchronous transfer mode (ATM).

The second part of the global telecommunications network is the customer premises network. Here, a distinction has to be made between premises networks for large business customers and premises networks for residential and small business customers. The networks for large business customers are very diverse since the telecommunications needs of these customers are also very diverse. General trends include the ever-growing amount of telecommunications traffic, the growing demand for flexibility (in terms of required bandwidth), the integration of voice and data applications and the integration of different data application solutions. The world is also facing the introduction of ATM in the form of ATM local area networks.

Compared with all these rapid changes, the customer premises networks for residentials and small businesses have remained relatively

Access to B-ISDN via Passive Optical Networks. Edited by U. Killat
© 1996 John Wiley & Sons Ltd

stable up to now. In most cases their main component is still the plain old telephone set, which is sometimes supplemented with a modem for data communications. However, rapid changes are also predicted for this part of the telecommunication network. One example is the personal computer that is becoming more and more a commodity, thereby creating the possibility and the need for inter-computer communication. A strongly growing group of residential users is already using services such as electronic mail, and is communicating via the quickly growing Internet. Another example is the television set, which is changing from a passive device into an instrument for interactive services such as video on demand, home shopping and video games.

The third part of the global telecommunications network is the part that connects the core network with the customer premises networks: the access network (AN). The AN can loosely be defined as the part of the telecommunications network that stretches out from the customer premises (i.e. from the border of the network operator's responsibility) up to the first local exchange (LEX). It is evident that when both the core network and the customer premises networks are changing rapidly, the link between these two networks cannot remain unchanged. Some trends are also clearly visible in the access network. With a smaller number of switches in the core network, the AN constitutes a larger portion of the total communications network and becomes, therefore, more important. By introducing more operation and maintenance capabilities in the AN, the costs of deploying this part of the network are reduced. Furthermore, general trends do not stop at the border of the AN; a greater capacity for the growing amount of telecommunications traffic, a higher level of integration and the digitalisation of the network also affect the AN. The trend of putting more optical fibre in the AN deserves special attention. For most of the trunk connections in the core network, the economic viability of optical fibre is already a fact. For the access network, the situation is, at present, less clear. This leads to different possible scenarios such as fibre-to-the-home (FTTH) or fibre-to-the-curb (FTTC), where the last drop from the curb to the home can be coax cable or twisted pair. Another aspect of ANs that might become important in the near future is the so called multi-operator access network. In a deregulated environment with competing network operators, several of these network operators might want to share the access network. This requires an access network that supports access to more than one core network, each owned by a different operator.

Since a clear distinction can be made between premises networks for large business customers and premises networks for small business and residential customers, it is convenient also to make a distinction between the AN for large business customers and the AN for small business and residential customers.

Access for large business must be highly flexible. Unused bandwidth that can be "returned" to the network operator, instead of leasing it permanently, translates directly into decreasing costs for the business. The access must also be highly reliable. For a lot of companies, one hour without their communication facilities is one hour without business, and therefore one hour without revenue. However, the focus of this book is not access for large business; the focus is the access network for residential and small business customers.

For these customers, the issue of costs is even more important than for large business. Costs that are too high for a specific telecommunications service or for a specific network concept are extremely prohibitive. Although it is not yet clear how telecommunications services will evolve in the future, it is obvious that at least some of these new services will require a high bandwidth and interactive facilities. Therefore the current analogue lines, or even ISDN connections and the separate CATV system with its one way traffic, are not enough to support all required future services. Flexibility with respect to new services, i.e. service transparency, and flexibility with respect to bandwidth allocation are also very important for the AN.

One technique that has the potential to fulfil the requirements for an AN for small business and residential customers is the combination of two powerful concepts, "ATM" and "PON", into a single system, "the ATM-based PON". A general outline of the ATM concept was presented in Section 1 "The ATM Technology" and an overview of PON technologies was given in Section 2 "Passive Optical Tree Networks". The combination of these two is the subject of this Section 3.

ATM technology offers flexibility in terms of service transparency and bandwidth allocation. It also provides a good set of capabilities for operating and maintaining end-to-end connections, thereby allowing a reduction in the operational costs of the network. With fibre in the access it becomes possible to offer high bandwidth services. The PON concept represents an optimum cost distribution for opto-electronic components. Furthermore, since the optical network itself does not contain active components, the system has a certain degree of inherent reliability.

Before explaining the technical concepts of an ATM PON in more detail, it is important to understand the ATM concept as applied to the access network in general. Therefore, a generic framework for the ATM-based access network is presented first. This framework is valid for any ATM-based access network, such as networks based on active multiplexers, concentrators, VP cross-connects, passive systems based on optical fibre techniques or systems based on coax techniques. The ATM PON will then be considered as a specific case in this generic framework.

3.1.2   A generic architecture for an ATM-based access network

The main function of the access network is to collect or multiplex traffic from several customers and offer it to the core network, and to distribute the traffic from the core network to the relevant customers. In the context of ATM, this means the multiplexing and demultiplexing of ATM connections; these ATM connections can be virtual path connections or virtual channel connections. The multiplexing and the demultiplexing functions constitute the kernel of the system, see Figure 3.1.1.

If the interfaces of the access system with the "outside world" are ATM based, e.g. ATM in SDH or ATM in PDH, transmission convergence to a specific physical layer and termination of this physical layer have to be performed. At the V_B side of the AN this is the "core interface function", and at the T_B side this is the "user interface function". From an evolutionary point of view, it must also be possible to connect customers with non-ATM-based interfaces with the AN. A good example is a customer with ISDN who does not require ATM. This is indicated by the T_X interface in Figure 3.1.1. In order to connect this customer, the AN must support some kind of interworking or mapping. Continuing with this example, the ISDN data can be mapped onto ATM cells and transported via the AN to an ATM switch or cross-connect. From there, the ATM cells containing the ISDN traffic can be routed via another mapping device to an ISDN switch. When interworking or mapping is performed, the ATM connection is terminated at that specific point.

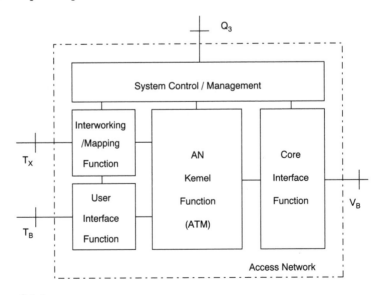

Figure 3.1.1
Functional architecture of the AN

An important part of the AN is the management system that performs functions such as fault management, performance management, configuration management and bandwidth management. All of these functions are located in the management plane of the B-ISDN protocol reference model. The Q_3 interface is the interface between the management system of the AN and the telecommunications management network (TMN).

A major objective for the AN is to keep itself transparent for user-to-network signalling. This means that the signalling end-points are located in the customer equipment and in the local exchange. All functions related to signalling, such as addressing, screening and charging are performed in the local exchange. Otherwise the functionality and the costs of the AN would increase dramatically.

One of the other major choices regarding the ATM layer of the access network is whether it should be similar to a VP cross-connect or to a VP/VC cross-connect. In a VP/VC-based AN, virtual channel links (and therefore implicitly also virtual path connections) are terminated and the AN has access to the VC sublayer. This means that functions which have to be performed in the AN, such as usage parameter control (UPC), operation and maintenance (OAM), routing of the ATM cells and cell header translation may also have to be performed at the VC level. For example, UPC and OAM at the VC level give better control possibilities of the access network, but require augmented hardware and software and put additional constraints on the exchange of control and management information between the core network and the access network. A VP/VC-based AN cannot be operated by a management system (i.e. via the management plane) only; additional functionality is required. Since virtual channel connections are set up on a call-by-call basis, the connection related information (e.g. bandwidth parameters) must also be transferred between the LEX and the AN on a call-by-call basis. This requires control plane functionality within the AN. On the other hand, in a VP-based AN, the virtual path connections are not terminated within the AN; only the virtual path links are terminated within the AN. This means that functions such as UPC, OAM, routing and cell header translation are performed at the VP level. This choice leads to a concept that is less complicated and therefore less expensive than the VP/VC-based AN.

3.1.3   The passive optical network: An example application of the generic architecture for the ATM-based access network

The generic functional architecture of the ATM-based access network can be applied to the specific example of an ATM-based passive optical network. The user interface functions and the core interface function are in principle the same as for any other ATM-based access network, e.g.

an active multiplexer. The main difference with other access systems is the kernel. For the ATM PON, the kernel is a cell transport system that runs over a passive optical tree-and-branch network. Specific ATM PON functions, as well as typical ATM layer functions, such as UPC, OAM flows and cell multiplexing, are performed in the kernel. These functions are discussed in more detail in the subsequent chapters.

The ATM PON system, as depicted in Figure 3.1.2, supports fibre-to-the-home (FTTH) and fibre-to-the-curb (FTTC). Both ATM-based and non-ATM-based interfaces can be offered, represented by T_B and T_X respectively. Different ATM-based interfaces can also be offered, such as ATM in SDH based on the STM-1 format or ATM in PDH based on 2 Mbit/s or 34 Mbit/s. The support of the physical layer has to be flexible, because new, cheaper interfaces might be standardised in the future. Non-ATM-based interfaces can also be supported, in which case the ATM connections are fully terminated in the optical network unit (ONU) and an interworking or mapping function in the ONU is required.

The optical distribution network (ODN) must comply with a certain set of specifications, e.g. in terms of maximum length between the optical line termination (OLT) and an ONU, maximum splitting ratio, maximum power loss and wavelength that has to be used. Another item that needs to be specified is whether the ODN is single- or double-fibre. The latter can be used for physically separate upstream and downstream traffic. The PON system itself consists of the OLT, a management system and the ONUs. It has to be engineered so that it

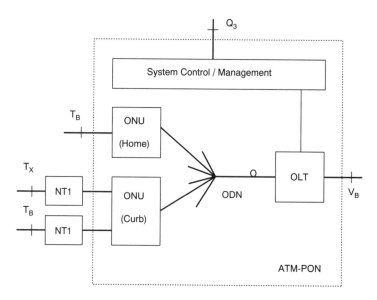

Figure 3.1.2
Architecture of an ATM PON: NT, Network Termination; ONU, Optical Network Unit; OLT, Optical Line Termination

can operate in co-existence with the specified ODN. It must be noted that an ODN may already be installed, e.g. if the current configuration of the AN consists of a narrowband PON. If the operator is considering an upgrade of the narrowband PON to an ATM PON, the preferable solution is to keep the ODN unchanged. So the specification of the ATM PON has to be made so that it complies with the (majority of) optical distribution networks that are currently being installed.

The OLT is the "head-end" of the PON system. It performs both the core interface function and a part of the AN kernel function as indicated in the generic functional architecture. Since the optical network has a tree-and-branch structure, it is convenient for certain functions to choose a master–slave type of control, instead of a distributed type of control. Following this approach, it is an obvious choice to locate the "master" functions in the OLT.

An ONU is located at each branch end of the optical network. It performs the user interface function and optionally an interworking or mapping function. It also performs part of the AN kernel function, or the "slave" part of the master–slave oriented control functions. The ONU can take the form of a so-called home ONU (ONU-H) or a curb ONU (ONU-C). The main functional difference between these two is the number of interfaces they support. In principle, the home ONU has a single interface, whereas the curb ONU has a number of customer interfaces.

For the cell transport system across the shared medium, a large number of viable options exists. Some of the basic options are, in principle, similar to the options that exist for narrowband PONs, as explained in Section 2. For example, for a specific ATM-based PON, a choice has to be made of the upstream multiplexing method. One can choose between TDMA, SCMA, CDMA, or WDMA, with trade-offs similar to those discussed in Part 2.

3.1.4 The BAF system: a specific implementation of an ATM-based PON

The BAF system is a specific system where choices have been made for all of the basic options for a cell transport system. In the remaining chapters of this book, these choices and their backgrounds are explained in detail. The basic concept behind the BAF system and its architecture are explained in the next chapter "System Architecture of the BAF system".

An important difference between an ATM PON and the narrowband PONs that are now being deployed is the data transmission speed. In general, ATM PONs are designed for a much higher speed (e.g. 155.52 Mbit/s, 622.04 Mbit/s or even 2.4 Gbit/s) than narrowband PONs, which have speeds of the order of 20–40 Mbit/s. The main

impact of the higher speed is on the physical layer aspects of the
TDMA system. This is explained in the next three chapters: Chapter 3.3
elaborates on the upstream synchronisation problem; Chapter 3.4
describes the transmission and reception of upstream ATM cells in a
burst mode; finally, Chapter 3.5 describes how the ONUs are virtually
kept at the same distance to guarantee that cells from different ONUs
do not collide in the upstream TDMA scheme. This is called the ranging
process.

Another important distinction between a narrowband and an ATM-
based PON is that the latter is able to support dynamic bandwidth
allocation. In a narrowband PON, a fixed number of channels is
transported via the bit transport system. In an ATM PON, cells can
be transported on a basis of "if-and-when-needed", thereby making
use of the flexibility in the ATM concept. A dedicated medium access
control (MAC) protocol is required to support this dynamic bandwidth
allocation, see Chapter 3.6. The subsequent Chapter 3.7 describes the
performance and the implementation of the MAC Protocol.

A general concern for shared medium systems is the privacy, security and authenticity of the information that is transported. In the
upstream direction, there is a possible threat of users impersonating
other customers connected to the same optical distribution network; in
the downstream direction, there is the threat of eavesdropping. Effective measures can be taken against these threats, see Chapter 3.8.

As explained before, a VP/VC-based access network cannot be operated by a management system only; additional functionality (control
plane functions) is required. However, when considering a VP-based
AN, it is possible to add such control plane functions. This creates the
possibility of adapting the size of the virtual path connections through
the AN on a call-by-call basis and therefore increasing the efficiency
of the available transport capacity within the PON system. A dedicated protocol between the switch and the AN, operating in the control
plane, is required. The so-called local exchange access network interaction protocol (LAIP) can be considered as the exponent of this control
plane functionality and is described in Chapter 3.9 "Interaction of a
PON-system with a B-ISDN Local Exchange". The management of the
BAF system is responsible for tasks that can be regarded as general
access network tasks, e.g. fault management, configuration management and bandwidth management. These tasks have to be performed in
the specific system environment of an ATM PON. How this is handled
in the BAF system is explained in Chapter 3.10 "Management aspects
of PONs".

Finally, the last Chapter 3.11, explains how the BAF demonstrator
will be integrated into two field trials in Europe, one in Turin (hosted
by CSELT) and the other one in Madrid (hosted by Telefonica).

3.2 SYSTEM ARCHITECTURE OF THE BAF SYSTEM

3.2.1 Introduction

The general architectural model of ATM PONs presented in Chapter 3.1 will now be adapted to the case of the BAF system. This chapter should be regarded as an example of translating general concepts into system design. The reader shall find a refinement of the general architecture to a suitable level of detail as a guideline for implementation and also an instance of this architecture implemented in the BAF system demonstrator. To avoid confusion between the system architecture and the particular demonstrator details, stemming from their parallel presentation as attempted in this chapter, the restrictions imposed by the demonstrator implementation on the BAF system concept are explicitly mentioned.

In this chapter the BAF system will be presented as a set of building blocks interconnected by interfaces. Along the lines established in the B-ISDN protocol reference model, user information is distinguished from management and control information. The flow of information is shown both vertically and horizontally. The vertical view establishes the dependency of functions residing in different levels of each building block while the horizontal flow allows one to identify the communication of peer protocol entities amongst the BAF system building blocks. The approach taken for the functional architecture is open in the sense that new functions can easily be introduced by identifying the layer in which they reside and their termination points.

A more detailed view on the set of functions performed within the BAF system is presented in a separate subchapter. The subject is presented in such a way that the reader is able to monitor the different functions performed on an ATM cell traversing the functional blocks of the BAF system. Special emphasis is put on the BAF system specific functions like ranging, encryption, VP translation and access control.

The internal U interface described in the next section deserves special attention. The U interface definition reveals the methods used to release the BAF system from the restrictions imposed by the well known standardised interfaces for user access and the core network and it reflects a number of selections made when the system was integrated.

The last but one subchapter is devoted to the problems arising from identifying connections in a signalling-less, broadcast, shared-medium configuration such as the ATM PON. The advantages of implementing a VP-based ATM PON are presented and efficient ways of exploiting the VPI field of the ATM cell header for addressing, routing and service identification are described.

The chapter concludes with a detailed discussion of the implementation aspects of the BAF system components.

Access to B-ISDN via Passive Optical Networks. Edited by U. Killat
© 1996 John Wiley & Sons Ltd

3.2.2 System configuration

The BAF system can be considered as a distributed access network operating on the VP sublayer of the B-ISDN protocol reference model as described in recommendation I.321 [2] by means of a passive optical distribution network (ODN). The optical line termination (OLT) is connected to a number of optical network units (ONU-C) for fibre-to-the-curb (FTTC) configurations, and/or optical network units (ONU-H) for fibre-to-the-home (FTTH) configurations. There are no active components in the ODN. All required management actions are performed by the operation system (OS). In Figure 3.2.1, the BAF system configuration is shown.

Both narrowband and broadband interfaces are provided on the T interface side, allowing also transparent transport of narrowband services on preallocated VPCs. In a general scenario, it is assumed that the broadband core network will inter-operate with the narrowband network. It is also possible to transport narrowband services between Tnb interfaces belonging to the same or different BAF systems, without the mediation of the narrowband network. The internetworking between the OS and the core broadband and narrowband networks is managed by means of an X interface. The X interface is not essential for showing the main features of the BAF system, and as such it has not been implemented in the BAF demonstrator. Instead the proprietary M interface has been used to manage the BAF system.

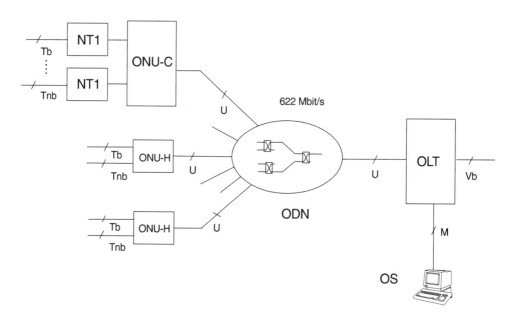

Figure 3.2.1
BAF system reference configuration

Provisions required to support CATV distributive services, using the same optical network, are accomplished at the U interface reference point by means of a different wavelength, together with wavelength selective components (see Section 2 of this book). The 1300 nm wavelength window is used for interactive services transport, whereas the 1550 nm window is left available for CATV distributive services transport. The motivation for supporting distributive services using a WDM technique is mainly for evolutionary reasons. More precisely, it is envisaged that, in the introductory stage of the ATM PONs for servicing small business and residential customers, the same optical infrastructure as the one already deployed for existing tree PONs will be used. Most of the existing tree PONs use different wavelengths for narrowband interactive and CATV distributive services. Moreover, the asymmetric nature and elevated bandwidth needed for high quality video distribution services might disturb the traffic profile of broadband interactive services.

The maximum bit rate the BAF system is able to support is 622 Mbit/s, both in the customer premises network to the core network direction (upstream) and in the core network to the customer premises network direction (downstream). The time division multiple access (TDMA) technique is used for the upstream traffic transport. A medium access control (MAC) protocol has been developed and implemented to resolve contention problems experienced when more than one ONU tries to make use of the common medium. The maximum splitting factor of the ODN is 32, and the maximum fibre length between the OLT and the ONUs is 10 km. In Chapter 3.4, the technical reasons for these constraints are presented in detail. The maximum number of T interfaces accommodated in the BAF system is 128. Due to technological constraints, only 81 of them are supported by the BAF demonstrator. The ONU-Hs are provided with one Tb (broadband T interface) and one Tnb (narrowband T interface). The ONU-Cs can be equipped with a maximum of 8 Tb and/or Tnb interfaces.

The U interface is located between the OLT and the ONUs. It is an internal interface and therefore it is not necessarily a standard one. It consists of two optical fibres, one for each transmission direction, at a bit rate of 622 Mbit/s. In the physical layer, a cell based frame structure composed of a three-octet preamble, supporting physical layer, OAM and MAC functions, and an ATM cell are used (see Chapter 3.2.4 for more details). Owing to the broadcast nature of a PON downstream, privacy and security are important issues. Encryption techniques on a per ONU basis operating in the physical layer have therefore been adopted in the downstream direction.

The external Vb interface is located between the OLT and the local exchange (LEX). The interface structure of the BAF system with the core network is one of the most important architectural issues of the access network. It is, however, very difficult to be concrete on the Vb interface definition, owing to lack of recommendations on this issue. The cell header layout conforms with ITU-T recommendation I.361 [3] for the NNI interface, that is, the VPI field covers 12 bits. At the physical level, the SDH standard is already available. With a total bit rate capability of 622 Mbit/s and SDH based transmission, three options exist conforming to ITU-T recommendations G.707-G.709 [8,9,10]: four physical interfaces transporting STM-1 frames, one physical interface transporting STM-4c frames and one physical interface transporting STM-4 frames. Since, at the moment, most of the equipment provides STM-1 based external interfaces, this option has been chosen for the BAF demonstrator. The higher levels of the Vb interface and, in particular the control functions, are still far away from standardisation.

The external Tb interface is readily assimilated to the standardised interface between the NT1 and NT2 network elements. The cell header layout conforms with ITU-T recommendation I.361 [3] for the UNI interface, adopting a 4-bit GFC field and an 8-bit VPI field. Since the BAF system is a VP based access network, policing is performed on a per VPC basis. Because the BAF system is an access network intended for small business and residential users, for the physical layer in general, and for FTTC configurations in particular, provision of inexpensive medium to low bit rate Tb interfaces seems an adequate approach. For the BAF demonstrator, 155 Mbit/s SDH optical interfaces have been implemented. In the same way as for the Vb interface, the only reason for implementing optical ones is that most of the equipment available supports 155 Mbit/s optical SDH based external interfaces.

The Tnb interface provides the possibility of transporting narrowband services transparently through the BAF broadband access network on preallocated VPCs. This interface supports unrestricted transparent transport of G.703 signals or unrestricted transport of G.704 frames at 2 Mbit/s, according to the relevant ITU-T recommendations [6,7].

The M interface interconnects the management workstation, where management functions and graphical user interface reside, with the OLT rack. The physical layer is based on a 10-BASE-2 Ethernet link running UDP/IP protocols.

The control bus that interconnects all the OLT boards and the OS has been implemented by means of a 10-BASE-2 Ethernet connection on the OLT backplane, running UDP/IP protocols. The control and management information between the OS and/or the OLT and the ONUs uses two different control channels: a so-called fast OAM channel and a so-called slow OAM channel. The fast OAM channel is used by the physical layer functions at the U interface, whose timing aspects are critical. It is implemented by means of some fields in the preamble attached to the ATM cells. Non-time critical OAM and management functions like the static allocation of VPI values, bandwidth allocation to VPCs (see Section 3.2.5), alarms, transfer of Tb SDH and Tnb management information to the OS, etc. use the slow OAM channel, implemented by means of ATM OAM cells inserted/extracted into/from the user ATM flow. The transfer of OAM information between the OS and the OLT is carried out via the Ethernet link mentioned above.

The Figure 3.2.3 depicts a protocol-wise view of the different functions of the BAF system and shows the points where the management and control information is inserted and extracted. It uses a very compact formalism defined and adopted in ETSI [1]; the reduced and simplified set of symbols sufficient for comprehension of the figure is explained in Figure 3.2.2. To avoid misunderstandings, it is worth noting that the reference points in the functional model and reference points in the reference configuration have different meanings. For instance, in ITU reference configurations S,T,U,V points are commonly used to indicate physical points, while functional model reference points identify interfaces between functions. In Figure 3.2.3 the BAF system is shown together with an ATM switch and a B-NT2 to indicate the interconnections to the rest of the communication network. This permits clear identification of the share of OAM/CTRL information processed by the BAF system and the share which is left untouched and traverses the system transparently.

The various abbreviations used in Figure 3.2.3 are described in Table 3.2.1.

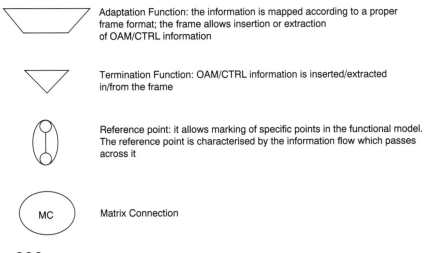

Adaptation Function: the information is mapped according to a proper frame format; the frame allows insertion or extraction of OAM/CTRL information

Termination Function: OAM/CTRL information is inserted/extracted in/from the frame

Reference point: it allows marking of specific points in the functional model. The reference point is characterised by the information flow which passes across it

MC   Matrix Connection

Figure 3.2.2
Formalism for functional models

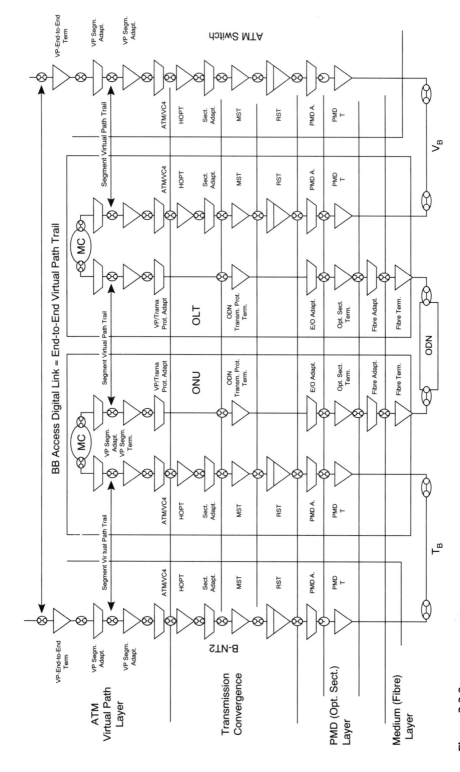

Figure 3.2.3
BAF system functional model

Table 3.2.1
List of abbreviations

Function abbreviation	Full function name
VP End-to-End Term.	Virtual Path End-to-End Termination
VP Segm. Adapt.	Virtual Path Segment Adaptation
VP Segm. Term.	Virtual Path Segment Termination
ATM/VC4	ATM/VC4 Adaptation
VP/Trans. Prot. Adapt.	Virtual Path/Transmission Protocol Adaptation
HOPT	Higher Order Path Termination
Sect. Adapt.	Section Adaptation
MST	Multiplex Section Termination
ODN Transm. Prot. Term.	ODN Transmission Protocol Termination
RST	Regenerator Section Termination
PMD A.	Physical Medium Dependent Adaptation
E/O Adapt.	Electro-Optical Adaptation
PMD T.	Physical Medium Dependent Termination
Opt.Sect.Term.	Optical Section Termination
Fibre Adapt.	Fibre Adaptation
Fibre Term.	Fibre Termination

The functionality can be partitioned into several layers and sublayers: the ATM virtual path layer containing all the functions involving VPI processing of the ATM cell; the transmission convergence sublayer (part of the physical layer) containing all the functions allowing the mapping of the ATM cells into a proper transport structure; the physical medium dependent sublayer (again part of the physical layer) containing all the functions needed to adapt the transport structure to the physical medium. In this sublayer the optical section includes all the optical functions relevant to transmission on a single wavelength. The remaining medium (fibre) layer, which is reached from the optical section of the PMD sublayer through an adaptation function which corresponds to wavelength division multiplexing, includes the functions common to all optical wavelengths (i.e. splitter/combiners and the structure of the optical distribution network).

Two main reference points can be identified in the upper part of the Figure 3.2.3 on the left and on the right. The ATM cells cross these points in both directions. These two reference points identify the end-to-end virtual path trail, which performs end-to-end transportation of a sequence of ATM cells. A special OAM cell flow, the so called end-to-end F4 flow [5], is added to each virtual path connection for performance monitoring purpose. The F4 flow is terminated in the proper termination blocks. A second F4 flow, called segment F4 flow [5], is terminated in the subsequent level of termination blocks. The purpose of this flow is to monitor the segment virtual path trail.

In the figure three segment virtual path trails are indicated: between B-NT2 and ONU, between ONU and OLT, and between OLT and the ATM Switch. The segment F4 flow is actually not implemented in the demonstrator to avoid an increase of complexity, but it is shown for completeness as it would be dealt with in a product type system. For more information on the F4 flows the reader is referred to Chapter 3.10.

Above the segment trails, in the ONU and in the OLT a matrix function is shown which effects the routing. As only ATM cells exist at this level, it can be deduced that routing is performed on the basis of the addressing information carried in the ATM cell (i.e. the VPI value).

The fact that ATM cells are transported in SDH frames outside the BAF system is represented in the figure by the sequence of termination and adaptation functions proper of SDH: high order path termination relevant to the ATM/VC4 adaptation, multiplexer section termination relevant to the (multiplexer) section adaptation, and regenerator section termination for the associated regenerator section adaptation. These functions handle what are called F3, F2 and F1 flows [5].

At this point the transport structure is complete and needs just to be adapted to the physical medium. Examples of such adaptation are the CMI encoding/decoding or the electrical/optical conversion functions. The termination blocks operate on physical layer monitoring information such as transmitter output levels and gain or equalisation properties of the receiver blocks.

Inside the BAF system the situation is different. When the ATM cells have been made available at VPI segment connection level (below the VP segment termination points), they are processed by the VP/transmission protocol adaptation function and the relevant OAM/CTRL information is inserted in the ODN transmission protocol termination point. Below these two blocks the data and OAM/CTRL information is mapped into a suitable frame structure (see Chapter 3.2.4). The information concerning the MAC, the encryption and the ranging is processed in this layer (see Chapter 3.2.3 for a description of these functions). General management information is also present in the frame structure in form of ATM cells; this flow of OAM information constitutes the internal F3 flow. When the direction of the information flow is bottom-up, the VP/transmission protocol adaptation function delivers to the upper layer a sequence of ATM cells "purified" from the general management, MAC, encryption, and ranging information.

The lower layers inside the BAF system perform the E/O adaptation and the optical section termination functions. Information concerning the optical flow is monitored and used to control the characteristics of the transmitters and receivers of the system. Bit timing is also performed in this sublayer.

The fibre adaptation function performs optical multiplexing; the fibre termination function has been included for completeness and it allows the identification of the optical wavelength.

3.2.3   BAF system specific functions

The way in which the user information is transported to and processed by the BAF system requires a set of specific functions to be implemented. Because such user information flow is the object of the analysis it is useful to define a function as a block which receives input user data, processes them on the basis of some processing information and produces output user data and output processing information for other blocks. Figure 3.2.4 depicts the relationship between these functions, adopting a three-dimensional representation.

Three planes can be distinguished in the figure, listed below in order of "depth":

- user plane: it contains the majority of the functional blocks, and is related to the processing of the information produced by and

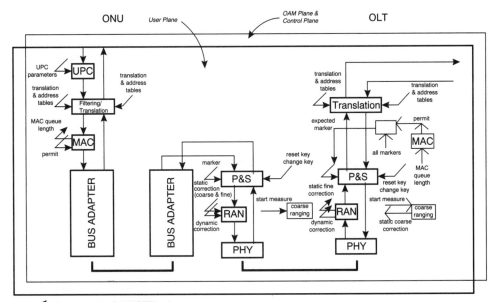

Figure 3.2.4
BAF system specific functions

addressed to the user; the user plane is partitioned into an ONU and an OLT plane,

- OAM plane: it is related to the transportation and processing of the operation and maintenance information needed to operate the system; examples of OAM functions are the ranging, privacy and security, system alarm collection and system performance monitoring. As it encompasses some less generic but BAF system specific functions the notion of the OAM plane has been chosen instead of the generic term management plane,

- control plane: it is related to the transportation and processing of the control information, typically regarding the call set-up and release as well as all the call parameters that have to be modified during the call; examples of control functions are the medium access control function and the translation functions and the usage parameter control function.

In the figure, for simplicity, the OAM plane and the control plane are shown as a single background plane, while the user plane has to be imagined in the foreground. All the functional blocks belong to one of the above-mentioned planes. Different arrows are used to show the information flow according to origin and destination of the information. The meaning of the arrows is also explained in the figure. It is worth noting that as the figure focuses mainly on the user information, it does not show how the OAM/CTRL information is transported, but simply assumes that such information is present where necessary. Not all of the user plane functions are described, but only the BAF system specific ones; for instance the V and T broadband and narrowband specific functions are not indicated.

At both ends of the functional block diagram data input and output consist of ATM cells. The cross-connection is performed on a VPI basis. It is useful to follow an ATM cell encountering the different functions shown in the block diagram to understand how the system works.

In the downstream direction the VPI is translated to allow flexible addressing of a specific T interface (see Chapter 3.2.5), and the cell is encrypted according to a mechanism implemented by the P&S block and according to the change-key and reset-key OAM information. The cell is passed to the PHY block performing the remaining physical medium and transmission convergence sublayer functions. In the ONU the cell is received by the PHY block which performs the usual physical medium and transmission convergence sublayer functions, resulting in the extraction of data and synchronisation signals. The P&S block decrypts the cell applying a mechanism which is the inverse of the one applied in the OLT. The cell is then broadcast to all T interfaces over the internal bus of the ONUs. Address filtering and VPI translation occur in the filtering/translation block which delivers to the subsequent blocks a stream of ATM cells addressed to that T interface with the corresponding VPI value in the user address space.

In the upstream direction the first function is the usage parameter control performed in the UPC block on the basis of connection related information such as peak bandwidth and duration of the burst at peak bandwidth. If the cell passes the check then it is relayed to the following blocks, otherwise it is discarded. This block is actually present only in T broadband interfaces: for narrowband connections the bandwidth is fixed. The absence of a UPC block in the OLT has to be highlighted. This is because the peer UPC block resides in the local exchange (LEX) and not in the OLT. In Chapter 3.9 an analysis of the exchange of control information between the LEX and ATM PONs, including the proper operation of UPC, is presented. The multiplexing performed in the upstream direction requires remapping of the user VPI values into a common BAF system addressing space to avoid mixing of user connections (see Chapter 3.2.5): this is performed in the filtering/translation block. A stream of ATM cells with converted VPI values is then delivered to the MAC block which stores all the cells in a queue. This block provides the control plane with the value of the length of the MAC queue; the control plane transports and uses this information in the OLT MAC block to establish the order in which the different T interfaces are allowed to transmit upstream. The information about the upstream transmission order is carried in permits. The figure shows how the permits are used. More information on MAC algorithm, permits, enhancements and performance can be found in Chapters 3.6 and 3.7. The MAC block delivers each ATM cell to the bus adaptor block in the proper time slot. The bus adaptor is in charge of transmitting and receiving the ATM cell across the ONU bus. The cell is then passed to the P&S block, where the cell is labelled with the ONU address in order to allow the OLT to recognise that a certain cell has really been transmitted by a certain ONU. This function is called origin checking.

The cell is then passed to the ranging block which introduces an equalising delay so that all the ONUs are continuously seen at the same distance from the OLT. This is necessary for an accurate synchronisation of the upstream transmission. The delay is made up of three contributions: static coarse, static fine and dynamic fine ranging delays, which as shown in the figure, are generated by different blocks in the OLT, on the basis of the timing of the received user data and of a special coarse ranging signal. It is important to point out that while coarse ranging is performed, data can be transmitted upstream with negligible performance degradation.

The cell is delivered to the PHY block which performs the physical medium and transmission convergence sublayer functions including the burst mode transmission. The arriving burst is processed by the PHY block at the OLT which performs all the physical medium and transmission convergence sublayer functions and delivers data and synchronisation signals to the subsequent blocks. In this PHY block two of the most critical functions for all TDMA PON systems are performed: the burst mode reception which is dealt with in detail in Chapter 3.4 and the bit and slot alignment to which Chapter 3.3 is dedicated.

The cell is then sent to the ranging block which produces the ranging information on the burst arrival time and realigns the data to the system clock. These functions, from the implementation point of view, are strictly related to the upstream synchronisation function. The P&S block performs the origin checking by comparing the arrival marker with the expected one. This is possible because the MAC block provides the information on the source of the expected cell.

The last BAF system specific function performed on the upstream cell is the translation of the VPI value, in order to make the BAF system internal addressing scheme independent from the external one (see Chapter 3.2.5). The flow of VPI translated ATM cells is then delivered to the V part of the OLT functional block which produces an ATM stream mapped into SDH transport modules (see Chapter 3.2.6).

3.2.4   U interface definition

As it has already been mentioned in Chapter 3.2.2, the internal U interface is located between the OLT and the ONUs, and it is a non-standard interface consisting of two optical fibres, one for each transmission direction, at a bit rate of 622 Mbit/s. The 1300 nm window is used in both directions for interactive services, whereas the 1550 nm is left for distributive services.

The U interface characteristics reflect the peculiarities of an ATM PON system in terms of system requirements and design challenges. For example, in PONs, different frame structures are used in the physical layer, closely related to the characteristics and physical layer constraints of the MAC protocol adopted. In the BAF system, a cell based frame structure composed of an ATM cell and a three-octet preamble, that supports physical layer, OAM and MAC functions, has been selected. Also, owing to the broadcast nature of a PON in downstream direction, privacy and security are important issues to be dealt with in the physical layer of the U interface. For this reason, encryption techniques on a per ONU basis are adopted for the downstream direction.

Another typical aspect of PONs is the burst mode character of the upstream traffic, owing to the large variations in amplitude and phase alignment of data coming from different ONUs. Upstream data scrambling is not needed, since the data coming from different ONUs are resynchronised to the system master clock using the sophisticated techniques described in Chapter 3.3. On the other hand, the ONUs are synchronised with the core network through the U interface, and so scrambling of downstream data is needed. Slot alignment of downstream data is acquired by means of the HEC mechanism described in ITU-T recommendation I.432 [4].

3.2.4.1 Downstream frame structure

The downstream frame structure at the U reference point consists of a continuous stream of BAF cells. Each BAF cell consists of a three-octet preamble and an ATM cell. The ATM cell format is defined according to ITU-T I.361 recommendation for the NNI interface. In Figure 3.2.5 each field is shown together with an indication of field site in bits.

For the downstream direction the preamble is split up into three subfields, each one devoted to a different set of functions:

- permit identifying the Tb or Tnb interface that is allowed to transmit information upstream.
- fast OAM commands including the identification of the destination ONU to which the fast OAM command is directed. (The fast OAM commands support physical layer functions like ranging and encryption).

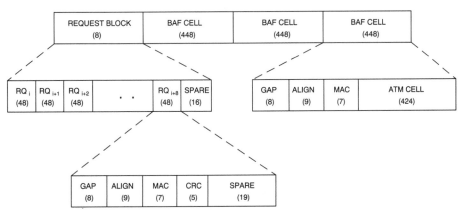

DOWNSTREAM FRAME STRUCTURE

UPSTREAM FRAME STRUCTURE

Figure 3.2.5
Frame structure

- CRC, that allows single error correction and multiple error detection for the 24-bit preamble.

The VPI field of the ATM cell is used to include the address information of the ONU to which the ATM cell is directed. This means that the BAF system adopts a particular addressing scheme at the internal U interface reference point realised by means of the VPI field which, for transparency reasons, leads to a VPI value translation functionality at both the Tb and Vb interface reference points.

3.2.4.2   Upstream frame structure

The upstream frame structure at the U interface reference point consists of a continuous stream of normal slots, empty slots, request blocks or static fine ranging slots (see again Figure 3.2.5). A normal slot consists of a three-octet preamble and an ATM cell. Again the ATM cell format is according to the recommendation for the NNI interface. The preamble layout includes a gap to avoid collisions between adjacent slots coming from different ONUs, a field used by the bit and slot alignment function, and a field "MAC" which contains the queue length associated with the particular Tb/Tnb interface that transmits the cell (see in Chapter 3.7 how the MAC protocol uses this information). Empty slots are received by the OLT during the system initialisation phase, when no cells are available to be transmitted upstream. A request block consists of nine consecutive requests of six octets each, plus two stuffing octets. Each request is issued by a different T interface. The request blocks are used by the MAC protocol to decrease the reaction time for the permit allocation mechanism. The request subfield of the request block looks similar to the preamble of a BAF cell but also includes a CRC field and a stuffing field.

The static fine ranging information consists of two octets, which are transmitted at the centre of an empty 56-octet long static fine ranging slot. The static fine ranging information subfield includes a field used by the bit and slot alignment functions, and a stuffing field for completion of the two-octet length.

The VPI field of the ATM cells indicates the address of the ONU from which the cell was issued. This information is used for performing authenticity functions (also known as origin checking) which protects the system against malicious use of the resources. Also, information about the physical Vb interface to which the cell is directed is provided within the VPI field of the ATM cell header. There are also VPI values reserved for internal OAM cells and ranging OAM cells. This means that, in the same way as in the downstream direction, the BAF system adopts a particular addressing scheme at the internal U interface reference point.

3.2.5   Connection identification

In shared medium topologies the identification of the source and the destination of a message is a central issue in the design of protocols. In connectionless LANs address identification is provided by special fields appended to each message while in connection oriented networks like B-ISDN, address indication is provided by a connection identifier which is assigned at the call set-up phase and uniquely characterises the communication session of specific users. Implicitly the connection identifier includes all addressing and routing information required by the network components. The ATM PON tree configuration illustrated in Figure 3.2.1 has an inherent broadcast nature: a downstream cell is received by all ONUs and an upstream cell can be forwarded to several V interfaces. Hence it is conceptually different from a normal ATM switch which has the capability to route cells according to their VCI/VPI value. Also, the ATM PON is not a terminating point for signalling information, which means that among others it ignores the values of the connection identifiers and furthermore their relations to the communicating parties, all established during the call set-up phase. Therefore connection identification, routing and addressing in ATM PONs are issues which look similar to the case of connectionless networks.

The ATM PON realisation for the BAF system is based on the VP concept. More precisely a number of its components, the ONU and the OLT, are able to interpret the VPI value of cells. At a first glance one would think that with such an approach the problem of connection identification is resolved if VPIs and their relations to communicating users are declared to the ATM PON during the establishment of a VPC. This is, however, only part of the truth. The problem is that the same VPI value may be used at different T and/or V interfaces for characterising different VPCs. In this respect the VPI value of a cell at the T or the V interface does not provide enough information for proper addressing and routing in the ATM PON. Another requirement stemming from the competition of the ONUs for the same medium is the need to distinguish among services. This would enable the development of access control mechanisms that recognise and accommodate services with diverse performance requirements. For example cells belonging to delay sensitive services may be required to access the common medium with priority over other less delay sensitive services. For these reasons VPI translation is provided at the entrance and the exit of the PON.

In a VP based ATM PON only the VPI field of the cell header is interpreted. Therefore the VPI value of all cells within an ATM PON is the logical resource or else the label determining the destination of the cell as well as the type of information it conveys. An *internal* algorithm should be applied for allocating VPIs to different VPCs. In general one can distinguish among static VPI value allocation and dynamic VPI value allocation. The static scheme requires that the pool of available VPI values is partitioned into subsets, one for each ONU. In this way the ONU, i.e. the destination of a downstream cell, is always unambiguously identified. An ONU cannot use the VPIs of a subset assigned to another ONU even in the case where its VPI values are exhausted. Since it is possible to have several V interfaces connecting the ATM PON to the LEX each subset is further divided into groups of VPIs equal to the number of V interfaces. In this way the VPI value of an upstream cell determines also the particular V interface which the cell should cross. Static allocation has the advantage of being simple, that is, there is no need for frequent internal communication between the ATM PON components because VPIs are recognised in the ONUs and the OLT according to a predefined rule. On the other hand, dynamic allocation of VPI values increases the number of available VPIs an ONU sees. However, it relies on an extended internal communication capability of the ATM PON components. More information on the dynamic allocation of VPI values in the ATM PON can be found in Chapter 3.9, where the exchange of information required between the PON and the LEX is discussed in detail.

For the BAF demonstrator only the concept of static allocation of VPI values has been used. In general, let a denote the maximum number of U interfaces, b the total number of T interfaces, c the maximum number of T interfaces connected to the same U interface, d the number of V interfaces connecting the ATM PON to the LEX, e the number of service classes recognised by the ATM PON access mechanism, and l the number of bits of the VPI field in the ATM PON. The number n of available VPI values in the BAF system equals

$$n = 2^l$$

For each U interface there are n/a available VPIs and for each T interface n/ac VPI values. The total number of VPI values available for each T interface and for a particular V interface becomes n/acd, and the number of available VPI values for each service class at a particular T interface using a particular V interface is $n/acde$. Applying the above to the BAF system architecture illustrated in Figure 3.2.6 for $b = 128$ this results in a total of 4096 VPI values ($l = 12$), in particular: 128 VPI values per U interface ($a = 32$), 16 VPI values per T interface ($b = 128$ and $c = 8$), out of which 4 VPI values point to one V interface. The V interface is then identified with a particular service class.

For reasons related to implementation complexity, usually a serious burden in building a system prototype, the BAF system demonstrator supports a maximum number of 81 T interfaces ($b = 81$) with 16 VPIs

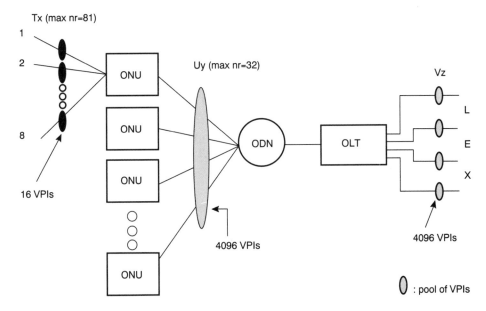

Figure 3.2.6
VPI values at the BAF system demonstrator interfaces

per T interface. The VPI value allocation scheme is based on the assumption that VPIs running over different T interfaces of the same ONU have different values, and that the V interface identity is only useful for the upstream direction. Hence the situation can be characterised as follows. In the downstream direction the first 5 bits of the 12-bit VPI are used to indicate the ONU address and only the first 4 of the remaining 7 bits of the VPI field are used for coding the BAF system VPI value of the VPC running over the T interface. In the upstream direction the situation is slightly different. Again the first 5 bits of the VPI field indicate the ONU, the following 2 bits determine the V interface ($d = 4$), and the first 4 of the remaining 5 bits encode the BAF system VPI value of the VPC running over the T interface.

At this point one should emphasise the fact that the specific VPI value allocation algorithm applied in the ATM PON system is an internal local loop matter which should be transparent for the core network. In other words the LEX should be able to allocate VPI values at the T and V interfaces independent of the specific needs of the access network for routing, addressing, and service identification functionality. This will guarantee the connectivity of the standard LEX interfaces to a number of diverse access systems, each one employing a different design philosophy. A solution that guarantees a properly functioning ATM PON being transparent to the LEX is to provide VPI translation functionality at the entrance and the exit of the PON; i.e., at the ONU and the OLT components. The inherent differences in the T and V interface characteristics and the independence of these interfaces as far as VPI value allocation is concerned provide another aspect of the VPI translation functionality. Note that at the T interface the UNI cell format is employed with a VPI field of 8 bits while at the V interface the NNI cell format uses an extended VPI field of 12 bits. With respect to the independence of the interfaces it is always possible that the same VPI value is assigned to two VPCs running over different T interfaces which are connected to the same ATM PON system. The same VPI value may also appear on two V interfaces connecting the ATM PON and the LEX.

The need for VPI translation becomes even more significant when the standardised VPI values of the T interface are considered. Table 3.2.2 presents the standardised values of the ATM cell header for different cell types as described in ITU-T recommendations I.361, I.432, I.610 and I.321. The zero VPI is reserved in each T interface for the support of several services including metasignalling and F3 OAM flows. To distinguish between the several T interfaces supported by the ATM PON it is necessary to translate such a zero VPI to another VPI value

Table 3.2.2
Standardised ATM cell header values; C: this bit is set to 1 in case congestion is experienced.

Cell Type	VPI	VCI	PT
Metasignalling	default: "0" any value	1	0C 0
General Broadcast Signalling	default: "0" any value	2	0C (0/1)
Point-to-Point Signalling	default: "0" any value	5	0C (0/1)
Segment OAM F4 Flow Cell	any value	3	0C 0
End-to-end OAM F4 Flow Cell	any value	4	0C 0
Segment OAM F5 Flow Cell	any value	Not "0"	100
End-to-end OAM F5 Flow Cell	any value	Not "0"	101
Resource Management Cell	any value	Not "0"	110
Idle Cell	0	0	000
Physical Layer OAM F1 cell	0	0	001
Physical Layer OAM F3 cell	0	0	100

which should be unique within the ATM PON. This means that in the ATM PON one VPI is reserved at each T interface or else, the number of available VPI values for user and control information transfer is reduced by the number of the supported T interfaces. It is important to mention that the same practice is followed also at the V interface where the zero VPI is always translated to another value. Since the ATM PON realises physical layer functionality the F3 cells concerned with the PL OAM functionality should be interpreted. To identify F3 cells the ONU and OLT components access the VCI and the PT field of all zero VPI cells (see Table 3.2.2). If the ATM PON is considered as a VP cross-connect, then segment OAM F4 flow cells should be interpreted and end-to-end OAM F4 flow cells should be monitored. In this case the ATM PON components should also access the VCI field of all cells and compare it with the standardised VCI values for segment and end-to-end OAM F4 flows (Table 3.2.2). Of course, this extra functionality puts additional complexity into implementation especially if this is provided by both the ONU and OLT components. The BAF system prototype is designed to provide F4 flow functionality in the OLT component.

Translation of VPI values is performed at the entrance and the exit of the ATM PON by virtue of VPI translation tables. The information maintained in these tables is represented graphically in Figure 3.2.7. In general an 8-bit VPIx at the Tx interface is translated to a 12-bit VPIy at the Uy interface (this is the ATM PON VPI value) which then becomes a VPIz at the Vz interface, where x is one of the T interfaces connected to the ATM PON, y one of the U interfaces of the ATM PON (note that the same U interface may be shared by several T interfaces) and z one of the V interfaces of the ATM PON and the LEX. The VPI value translation tables can be updated dynamically, each time a new VPI

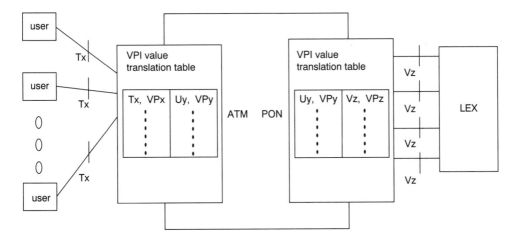

Figure 3.2.7
VPI value translation tables

is established by the LEX which provides signalling capability. This requires the existence of an LEX access network interaction protocol (LAIP) over the V interface which must be responsible for transferring to the ATM PON the information required for the construction of its tables. A detailed presentation of the LAIP protocol which has actually a broader scope than VPI translation is presented in Chapter 3.9. By virtue of the LAIP protocol, the VPI translation functionality together with the internal allocation of VPI values allows for flexible support of point-to-multipoint downstream services such as, for example, video on demand (VOD). The general idea is that the VPI value of a downstream multicast VOD cell at the V interface may be translated to a number of VPI values at different T interfaces equal to the number of the VOD subscribers. This would require an enhanced OLT able to provide multiple copies of VOD cells.

3.2.6 System components

3.2.6.1 OLT

The OLT functions are represented in Figure 3.2.8. In the downstream direction, the data coming from the LEX through the four 155 Mbit/s optical Vb interfaces based on SDH STM-1 frames are extracted, and the four ATM flows are transferred to the ATM mux adapter block. The main functions performed at the SDH to ATM termination points are frame acquisition, cell delineation and HEC verification. The reference clocks for system synchronisation are extracted and the Vb interface

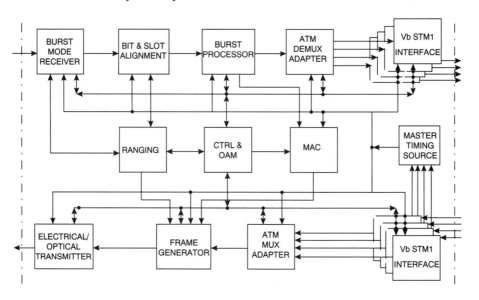

Figure 3.2.8
OLT functional block diagram

physical layer performance monitoring functions are performed at this point. The ATM mux adapter block performs the translation of the VPI values into the BAF system internal ones. Also, the insertion of control information towards the ONUs in the form of slow OAM cells is realised. Then the four 155 Mbit/s ATM flows are multiplexed into one single 622 Mbit/s ATM flow and delivered to the frame generator block. Since there are four clock sources, one from each Vb interface, this block synchronises the four ATM flows to the master clock. In the frame generator block the ATM flow is encrypted on a per ONU basis, and the physical layer preamble is added to each ATM cell. The preamble is filled with the MAC information provided by the MAC block, the fast OAM information provided by the ranging and control blocks and the CRC which is internally calculated. The whole data flow is scrambled. The HEC field of the ATM cell is recalculated over the scrambled data in order to enable the cell delineation functions on the ONU side to work properly, and then the cell is passed to the electrical/optical transmitter block, where the cells' flow is converted into optical power injected into the fibre.

In the upstream direction, the data coming from the different ONUs through the 622 Mbit/s optical U interface are received by the burst mode receiver, where the optical power is converted into electrical signals and transferred to the bit and slot alignment block. Also the low-frequency low-level coarse ranging signal is detected and passed to the ranging block. The bit and slot alignment block performs the upstream data phase alignment and frame synchronisation with the system clock. This is one of the most critical tasks in the system. Also some control signals concerning dynamic ranging are provided to the ranging block. The data flow is then transferred to the burst processor block, where the MAC information is extracted and the preamble is discarded. In this block, U interface physical layer performance monitoring functions are also carried out. The ATM flow is then transferred to the ATM Demux Adapter block. The ATM Demux Adapter block is responsible for extracting the slow OAM cells to be directed to the OAM plane, re-routing the incoming ATM cells to the appropriate Vb interface and translation of the internal BAF VPI values into the ones agreed with the LEX on each Vb interface. In the Vb interface block, the four ATM cell streams are inserted into STM-1 frames, converted into optical power and injected into the fibres. Other functions performed in this block are HEC generation for ATM cells and SDH OAM overhead information insertion.

It is worthwhile mentioning the following other functions.

• Loopback capabilities are also included in most of the data path blocks for subsystem and system integration aid.

- There is a master timing source block which receives reference clocks from the four Vb interfaces, and selects one of them as the master clock of the system. The BAF system works synchronously with respect to the LEX.

- There is also a control block where the MAC, ranging and encryption protocols reside, which is responsible for intercommunication with the OS.

- ATM layer performance monitoring is also done in the ATM mux/demux adapter blocks.

- Authentication is performed in the upstream burst processor block. The MAC block provides the burst processor block with information stating which ONU is allowed to transmit in which slot. In the burst processor block the origin of the received slot is checked against this information.

3.2.6.2   ONU

An ONU-H provides a single Tb interface plus a single Tnb interface and it is used for FTTH configurations; an ONU-C provides multiple Tb and Tnb interfaces, and it is used for FTTC configurations. The ONU functions are represented in Figure 3.2.9. The main functions performed by the different blocks are outlined below, starting from the description of the downstream path (from U interface to Tb, Tnb interface).

The optical downstream signal is received and converted into an electrical signal (O/E); the clock is extracted and data regenerated (Ck rec); data is parallelised (S/P) and all the master timing signals are generated (MTS). In the frame processor (FP) the data are byte and cell aligned according to ITU-T I.432 [4] and then descrambled; MAC information as well as (fast) OAM information (both carried in the preamble) is extracted, checked for errors and then passed to the MAC block and to the OAM block, respectively. The ATM cell is decrypted according to the rules established by the decrypt block, (slow) OAM cells are extracted from the downstream flow and sent again to the OAM block.

User cells are broadcast on the ATM bus to all T interfaces. The PBIA down block receives the ATM cells from the ATM bus and performs functions belonging to the ATM layer (VPI translation and cell demultiplexing) selecting all the cells addressed to the T interface. In the case of a T broadband interface all the ATM transmission convergence sublayer functions are then performed in the ATM-TC down block (cell rate decoupling and HEC generation) and the TTF down block (transmission frame adaptation and generation). The resulting ATM cells mapped into STM-1 frames are then passed to the Tx block which performs the physical medium sublayer functions (P/S conversion and

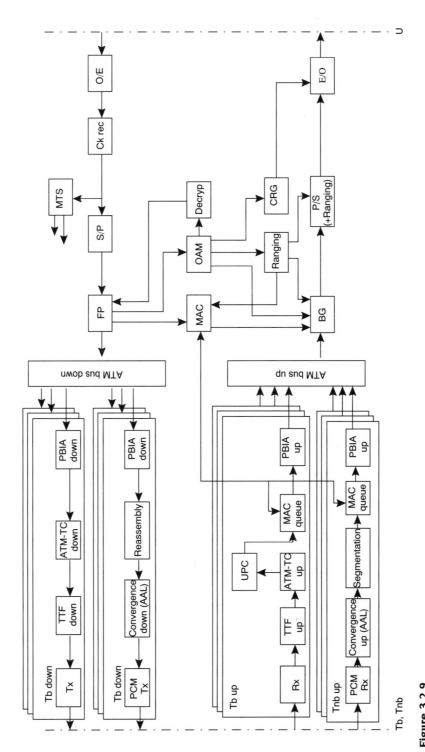

Figure 3.2.9
ONU functional block diagram

E/O conversion). In the case of a T narrowband interface, the ATM cells selected by the PBIA down block are passed to the remaining blocks to perform frame reassembling and AAL convergence functions (compensation of cell delay variation and timing recovery). The data are then passed to the PCM Tx block for HDB3 encoding and transmission.

In the upstream direction, both Tb and Tnb data have to be processed in order to produce ATM cells to be placed into the MAC queue. From that point on, the ATM cells undergo the same processing. In the case of a Tb interface, the STM-1 frame is received and the clock is extracted (Rx block); the frame is then passed to the TTF up block for frame recovery and frame adaptation; the ATM-TC up block then performs the cell delineation, HEC verification and cell rate decoupling. The cells are delivered to the UPC block, which performs the policing function, and then they enter the MAC queue. In the case of a Tnb interface, the incoming 2 Mbps stream (in HDB3 encoded G.703 [6] format) is received and the clock is extracted. The AAL convergence sublayer functions are then performed (AAL convergence up block) and data are segmented resulting in the generation of an ATM cell stream (segmentation block) which is then passed to the MAC queue block.

Cells are extracted from the MAC queue on reception of a proper command of the MAC block, and passed to the PBIA up block which performs a fine synchronisation of the cells to allow an error-free multiplexing on the upstream ATM bus. In front of each cell the information on the MAC queue length is attached. The composite data are received by the burst generator (BG) which attaches the physical layer overhead necessary for bit/byte alignment in the OLT and performs part of the ranging adjustment. The data are then serialised while fine ranging adjustment is performed; the serial data are then passed to the E/O block for electrical-optical conversion and transmission onto the network.

Other important functions are:

- The upstream OAM information is generated by the OAM block and inserted by the BG block in the upstream flow on reception of a proper MAC command.

- To increase the reaction speed of the MAC protocol, special blocks made up of nine requests can be transmitted upstream by the ONUs in place of a cell (see Chapters 3.6 and 3.7 for more information on the MAC protocol and request blocks). Each request contains the length information for a certain MAC queue and is therefore relevant to one T interface. Requests coming from different ONUs can be assembled "on the fly" in a single request block.

- The coarse ranging procedure requires a low-level low-frequency tone to be generated in synchronism with the reception of a proper command sent by the OLT. This is accomplished by the CRG block which is activated with a precise timing signal by the OAM block.

- The equalisation of the round trip delay necessary for the synchronisation of the TDMA scheme used in the upstream direction is controlled by the ranging block which acts on the MAC block to delay the instant in which the "extract cell from MAC queue" command is issued, on the BG block to insert a fixed number of byte clock delay cycles, and on the P/S block to perform the finest correction at bit level.

3.2.6.3 ODN

The signals exchanged between the OLT and the ONUs travel over a double tree structured PON. The network is depicted in the Figure 3.2.10. The main characteristics of an ODN are:

- distance range,
- splitting/combining factor: i.e. number of "leaves" of each tree.

From the optical point of view the following aspects play an important role:

- attenuation,
- attenuation unbalance,
- reflections.

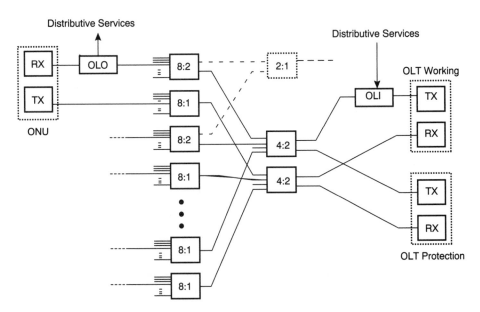

Figure 3.2.10
Optical distribution network; OLO, optical line outlet; OLI, optical line inlet

The attenuation depends on the distance and on the splitting/combining factor. It is worth noting that combiners and splitters are the same type of device, and introduce the same attenuation independently on the travelling direction of the light. The attenuation unbalance depends on the maximum and minimum distance as well as on all the non-uniformities of the splitters and combiners. The attenuation unbalance has an impact on the dynamic range of the receivers and should be maintained as low possible. Reflections can provoke a degradation of the performance of the system: for instance, upstream optical power could be reflected back towards downstream receivers causing a sensitivity impairment.

The characteristics of the ODN of the BAF system are briefly summarised below:

- distance range: 0–10 km
- splitting factor: 32
- optical attenuation: 15–30 dB
- wavelength: 1300 nm window

Some architectural aspects of the PON deserve to be mentioned. The same optical wavelength in the upstream and downstream directions is chosen to leave an optical window for other services. As a matter of fact one of the simplest methods of allowing the co-existence of different services is to employ different optical wavelengths and use WDMs (OLO, OLI in the figure). Depending on the characteristics of the optical paths the WDMs can be positioned in the upstream or downstream direction.

In general different splitting factors are associated with different services. The problem can be solved rather easily by adopting multiple levels of splitting factor and by inserting the service with the lower splitting factor at the proper level. An example for a 1:16 splitting factor is shown in the figure.

Two separate trees can be identified in the network: one for each direction. The downstream and upstream signals do not mix together allowing the use of the same optical wavelength without too many constraints on component reflections. Lasers with no particular requirement on optical spectrum can advantageously be used.

The double fibre to the first splitting level allows a 3 dB gain, because it permits the removal of the internal RX-TX combiner in the OLT. Whenever a system component serves a "large" number of users, it is necessary to foresee a protection path in the architecture of the system. This concept, linked to the double fibre concept, explains the structure of the optical network on the OLT side.

REFERENCES

[1]   ETSI TM3 DTR/TM-3007 *B-ISDN Access*, 1995
[2]   ITU-T (CCITT) recommendation I.321 *B-ISDN Protocol Reference Model and its Application*, 1991
[3]   ITU-T (CCITT) recommendation I.361 *B-ISDN ATM Layer Specification*, 1993
[4]   ITU-T (CCITT) recommendation I.432 *B-ISDN User–Network Interface Physical Layer Specification*, 1993
[5]   ITU-T (CCITT) recommendation I.610 *ISDN Operation and Maintenance, Principles and Functions*, 1993
[6]   ITU-T (CCITT) recommendations G.703 *Physical/Electrical Characteristics of Hierarchical Digital Interfaces*, 1991
[7]   ITU-T (CCITT) recommendation G.704 *Synchronous Frame Structures Used at Primary and Secondary Hierarchical Levels*, 1991
[8]   ITU-T (CCITT) recommendation G.707 *Synchronous Digital Hierarchy Bit Rates*, 1991/1993
[9]   ITU-T (CCITT) recommendation G.708 *Network Node Interface for the Synchronous Digital Hierarchy*, 1991/1993
[10] ITU-T (CCITT) recommendation G.709 *Synchronous Multiplexing Structure*, 1991/1993

3.3 THE UPSTREAM SYNCHRONISATION PROBLEM

3.3.1 Introduction to burst mode synchronisation

ATM cell transmission in the upstream direction on the ATM PON network can use TDMA interleaving of cells. In this case, the transmission is burst mode and a ranging procedure must ensure that cell collisions do not occur by controlling the transmission instant of each ONU. The accuracy with which this function can be performed in the BAF system is to within approximately 2 bits, therefore the cells arrive at the OLT in a near continuous stream, separated by gaps of several bits.

The gaps between received cells at the OLT will be a non-integer number of bits in duration, and this largely rules out conventional techniques for synchronisation. The phase-lock loop (PLL) has been widely used in synchronisers, since this device allows the clock to be recovered from a data signal. However, existing PLLs require a certain time to acquire phase lock, and in this sense they are too slow. Moreover, PLLs show a significant amount of phase error while the loop is acquiring lock [3]. The synchroniser for the OLT must acquire phase lock in the preamble of each cell, since the phase of each cell is unknown. Methods have been proposed for high-speed transmission systems that lock on the first transition in the data preamble; these have been reported by [4,5,6].

In an ATM PON network, the bit and cell synchronisation at the optical line termination can be performed by phase aligning each arriving ATM cell with the master clock at the OLT. Since the ONU clocks are frequency locked to the OLT master clock, then the upstream data will be frequency locked to the OLT master clock. Phase aligning each cell with the master clock will therefore allow the master clock to be used for data retiming, and for generation of lower speed clocks.

Phase alignment must be performed on a cell by cell basis, since each cell is separated by a gap of non-integer bit length duration. At the end of each cell, the phase aligning circuit must be reset, and this signal is directly derived from the master clock. Ranging ensures that the resets occur during the gaps between cells, by delaying the transmission of upstream cells appropriately. One method used for phase alignment is based on the digital technique of unique word detection, first proposed over 25 years ago [1]. This technique has been substantially developed for the BAF system [2] and is outlined in Chapter 3.3.2. As an alternative in Chapter 3.3.3 an analogue solution based on low-cost silicon bipolar circuits is presented.

3.3.2 Digital technique for phase alignment

A keyword (or: unique word) transmitted in each cell preamble is sampled a number of times per bit. From those samples that allow the

Access to B-ISDN via Passive Optical Networks. Edited by U. Killat
© 1996 John Wiley & Sons Ltd

keyword to be recognised, through proper latching of the keyword, a clock phase is selected for latching the remaining data of the ATM cell. The data can then be aligned with the master clock at the OLT, since the phase relation between each chosen clock phase and the master clock is known.

Once bit alignment has been performed, it is necessary to perform cell alignment, since the input phase variation of each cell is greater than one bit. The output of the synchroniser is demultiplexed, hence this demultiplexer needs to be aligned to each cell so that the cell header bytes are demultiplexed wholly. The structure of the synchroniser can be of a feed-forward design, allowing the use of the same keyword for both bit and cell alignment, since the keyword is not corrupted. Cell alignment can therefore be performed by simple recognition of the keyword in the data stream.

In the following subchapters, the structure of the synchroniser is further described. It can be broken down into a number of blocks: (i) the delay-lock loop, (ii) the aligner, (iii) the comparator, (iv) the selector, (v) the switch, (vi) the demultiplexer, (vii) the ranging counter, as well as the control logic. This is illustrated in Figure 3.3.1.

A keyword in the ATM cell preamble is sampled at a multiple of the data frequency. Generation of a multiphase clock from a delay-lock loop facilitates this. Thus eight data channels, representing data sampled with each phase of the clock, are then synchronised with a master clock. Metastable states are also resolved in the aligner.

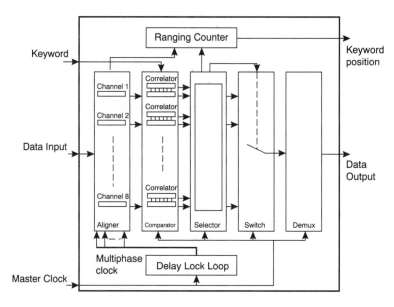

Figure 3.3.1
Schematic of a digital burst mode synchroniser

The comparator searches the data for the keyword, resulting in "hits" i.e. recognition of the keyword, "misses" i.e. no recognition of the keyword, and "mishits" i.e. incorrect recognition of the keyword. The selector finds the centre of the "hits" representing the optimum phase of the sampling clock. The data is then routed via the switch to the demultiplexer, where further cell alignment is performed. The output is byte synchronous. The ranging counter provides information as to the detected position of the keyword.

3.3.2.1 *The delay-lock loop*

The keyword in the cell preamble is sampled a number of times per bit. One way of performing this is to generate a multiphase clock, thereby sampling the keyword with each clock phase. To generate a multiphase clock, a tapped delay line can be used, whereby each stage of the delay line represents a phase shift and the total delay of the line represents one bit period. An eight-stage tapped delay line is illustrated in Figure 3.3.2, represented by a chain of differential driver stages. The eight differential clock phases generated from the tapped delay line are illustrated in Figure 3.3.3. These are spread fairly evenly over a period of 1.6 ns, corresponding to an operating frequency of the synchroniser of 622 Mbit/s.

Implementing such delay lines on chip will give rise to delay variations due to a number of effects. These are process variations, temperature, power supply variations etc. In total, these variations may account for a 2:1 variation in delay. One method that can be used to dynamically tune the delay line is to incorporate it in a delay-lock loop (DLL). This consists of three elements, (i) a digital phase detector, (ii) a loop filter and (iii) a voltage or current controlled delay.

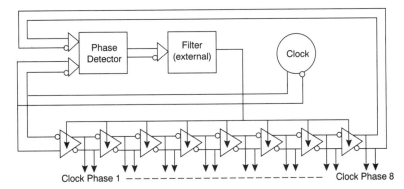

Figure 3.3.2
Schematic of the delay-lock loop (DLL). A controllable tapped delay line constructed from controllable differential driver elements is shown. In addition, a digital phase detector and an active filter complete the DLL. The clock is provided externally

Delay Line Taps

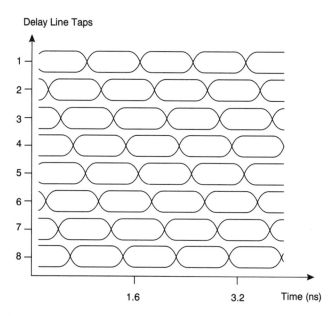

Figure 3.3.3
Eight differential clock cycles, separated by 200 ps, are evenly spread over a clock
cycle of 1.6 ns

The digital phase detector compares the clock signal applied at the
input of a controllable delay with the output of a controllable delay,
and when the two are in phase, the delay represents a phase shift of one
period. This is the phase-lock condition, where the differential output of
the digital phase detector gives no correction signal, and its differential
digital output acquires a mean value as shown in Figure 3.3.4.

The controllable delay line can be implemented on chip. For a semi-
custom chip, some customisation of the cells is necessary to implement
such a device. Each customised delay cell must give an adequate
tuning range. For one particular implementation of the delay cell, the
bandwidth is reduced when the delay is extended, since current is
reduced in the cell switching circuitry. This means that it is preferable
to pass the clock signal through the delay cell, and thus generate a
multiphase clock signal. Transmitting the data signal through such
a delay line would introduce severe intersymbol interference on the
waveform when the bandwidth of the delay cell is not sufficiently high.

The closed-loop response of the DLL is illustrated in Figure 3.3.4.
The input control current for the tapped delay line ramps up until the
total delay of the delay line is reduced to one clock cycle. Here, the
correction signal from the phase detector is reduced to a mean value
and the loop settles.

The accuracy by which the controllable delay line is tuned to one
period of the clock signal is not critical in DLLs since no accumulation

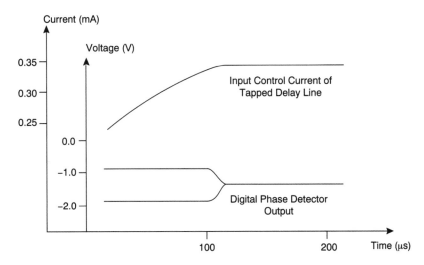

Figure 3.3.4
Closed loop response of the delay-lock loop. The input control current to the delay line ramps up until the phase-lock condition, where the differential ECL output of the digital phase detector gives no correction signal

of phase error occurs. If the controllable delay line fed back its inverted output to its input, forming a ring oscillator, a controllable oscillator would result. Thus a phase-lock loop (PLL) could be constructed around the oscillator. The drawback is that accumulation of phase error would occur for the PLL. Thus a DLL is preferable to a PLL.

3.3.2.2   The aligner

The multiphase clock is used for sampling the data in a number of phases. Most phases will allow proper latching of the data. However, for some phases, the clock-phase margin of the sampling bistable circuit will be insufficient, and metastability may arise [7,8]. The differential output of a bistable circuit in a metastable condition may be at a mean level as illustrated in Figure 3.3.5. In time, the output will diverge to the normal logic levels. The time that this takes for a given device is dependent upon the finesse to which the voltage levels internal to the device's switching circuit are balanced, at the moment the bistable circuit is clocked. If the voltage difference internal to the device's switching circuit is of the order of nanovolts, the bistable circuits differential output may take several microseconds to resolve.

If an unresolved signal is clocked by further stages of a shift register, the differential voltage of the signal will be increased by the amplification of each stage of the register until the signal at the output of the shift register is fully resolved. If the input phase variation of the received data is uniformly distributed, then the probability can be predicted of

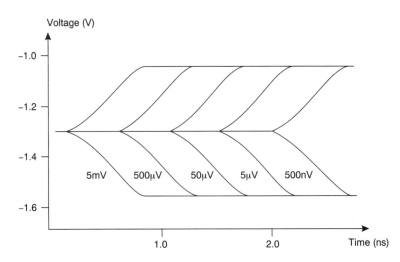

Figure 3.3.5
Metastable state resolution of a bistable circuit. The output differential voltage
divergence with time, for given input differential voltages (5 mV to 500 nV), at the
moment the bistable circuit is clocked

data being clocked with less than the required differential input voltage
at the sampling bistable circuit, for metastability resolution in a given
time. Therefore, this determines the length of shift register required to
ensure that metastable states are resolved at its output at a probability
less than the cell loss ratio, preventing metastable states entering further
parts of the synchroniser and potentially jeopardising its operation.

The aligner must take all data streams corresponding to each phase
of the clock, and align them to the master clock, to allow synchronous
operation of the following circuitry. Since the relation between the
master clock and each clock phase is fixed, this can be achieved fairly
easily.

3.3.2.3   The comparator

The comparator compares the latched data resulting from sampling at
each clock phase with that of a predefined keyword. This is performed
with a digital correlator for each channel, whereby each correlator
allows for one bit error. For an n-bit keyword, the correlator searches
for n-1-bit correlations. This allows recognition of the keyword even
if it is corrupted by a bit error, thereby allowing cell loss ratios better
than 10^{-15} to be achieved for bit error rates of 10^{-9}. This is determined
from the probability of two bit errors occurring in the guard band or
keyword, the failure condition.

The comparator can also be designed so that it allows programma-
bility of the keyword, and the length of the keyword. Furthermore,

masking of certain bits can be implemented, to allow the synchroniser to ignore regularly distorted bits. For example, the first bit of
the preamble may always be distorted due to the burst mode receiver.

3.3.2.4 The selector

The selector processes the correlator outputs and must choose a suitable
data channel from this information. When the correlator recognises the
keyword in the data that has been sampled with a suitable clock phase,
the correlator registers a hit. When the keyword is in proper alignment
with the installed keyword, multiple hits will occur, corresponding to
those clock phases that fall within the open part of the data eye. For
an eight-channel synchroniser as illustrated in Figure 3.3.1, there may
be eight or even nine hits. Nine hits may result in an eight-channel
device when the mth and $m + 8$th clock phase falls exactly on the data
transition, and the metastability resolves in favour of recognition of the
keyword in both cases.

When the data sequence is sampled and the keyword is in improper
alignment, it may also be possible to acquire correlation with the
installed keyword, and this is referred to as a mishit. It must be ensured
that mishits do not corrupt the operation of the selector, to prove the
technique is suitable for data synchronisation.

Mishits can occur owing to the autocorrelation of the keyword with a
shifted keyword, in the presence of bit errors. This may happen when
the data sequence containing the keyword is not fully clocked into
the comparator. For an n-1-bit correlator, we must ensure that there
is a mismatch of at least three bits when the keyword is in improper
alignment, to prevent mishits. The eight-bit keyword 11100101 satisfies
this requirement.

Furthermore, autocorrelation can occur between the keyword, and
a distorted version of itself, again when the keyword is in improper
alignment. This problem is illustrated in Figure 3.3.6. With the addition of a bit error, the bistable circuit that takes a clock phase occuring
near to the data transition can latch either state, depending on the arbitrary pulse width distortion on that transition. This problem occurs at
data transitions in the keyword, and can be circumvented by a suitable choice of keyword. Keywords with combinations of successive
bits cannot be affected in the same way, even in the presence of a bit
error. The choice of the keyword is therefore determined not only by
the autocorrelation properties of bits, but also the relative positions of
consecutive bits and transitions. The keyword 11100101 satisfies both
these issues.

When the keyword is in proper alignment, a number of hits will
occur and the selector will locate the centre of the group of hits. The
centre hit from the group should correspond to the optimum clock

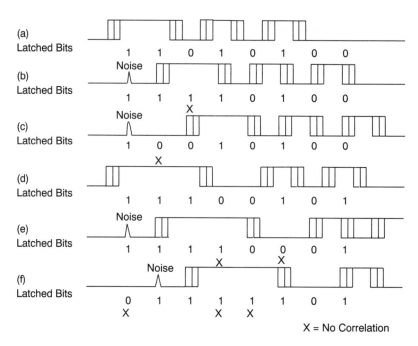

Figure 3.3.6
Latched states of a decision circuit, when the input data shows pulse width
distortion, and noise. (a) Keyword 11010100 in proper alignment, a "hit"; (b) and
(c) keyword 11010100 in improper alignment. Latched data gives 7-bit correlation, this is a "mishit"; (d) keyword 11100101 in proper alignment, a "hit"; (e) and
(f) keyword 11100101 in improper alignment. No "mishits" occur, since there is no
7-bit correlation, even with a noise spike

phase for those data to be retimed with. On the periphery of such a
group of hits, there may be some mishits, especially if there are multiple
transitions on the data bits at the output of the optical receiver, due
to ringing etc. Secondly, metastable effects may also result in mishits
occurring around the group of hits.

To prevent jeopardising the operation of the selector in the presence
of mishits, the selector can be configured to ignore singular or pairs
of hits or mishits, since when the keyword is in proper alignment,
at least three hits should result. Thus mishits will be deleted if they
occur in isolation or as an isolated pair. If the mishits are on adjacent
channels to the hits, they will represent an offset in locating the centre
of the group of hits. However, with two mishits and only three hits,
the centre channel is still a hit.

Incorrect operation can therefore only occur if three successive
mishits occur, and this can be predicted to occur only when the input
data waveform contains multiple transitions of each bit that extend for
over one quarter of the bit period. Pulse width distortion on the data
waveform does not affect the selector in the same way, and it can be

predicted that the synchroniser can tolerate up to approximately 70% eye closure, sufficient to give three hits.

3.3.2.5   The switch

Once the selector has chosen the channel corresponding to the optimum clock phase, then the switch simply routes this channel to the serial output. Since the configuration of the synchroniser can be feed-forward, the preamble can be output, undistorted.

3.3.2.6   The demultiplexer

ATM cell alignment must also be performed on a cell by cell basis, since the phase variation between cells is greater than one bit. By recognition of the same keyword used for bit alignment in the preamble, then the demultiplexer can be aligned so that the first bit of each cell can always occur on the same parallel output pin. This ensures a fully synchronous parallel output of data bytes.

3.3.2.7   Ranging counter

Ranging is the method by which upstream bursts are roughly synchronised; so cells optically interleave at the passive splitting point, thus avoiding collisions. Ranging is subdivided into three stages, coarse, static fine, and dynamic fine ranging. For static fine and dynamic fine ranging, a ranging counter measures the time between the start of the slot, and the arrival of the cell preamble within the slot. For static fine ranging , this duration can be as much as several hundred bits. The ranging counter can be located within the synchroniser, thus giving an accurate measurement to within an 1/8th of a bit, achieved by combining the bit counter output with an indication of the clock phase selected for each keyword.

3.3.3   A bit synchroniser using analogue techniques

In the bit synchroniser presented here, an analogue phase detector is used to process the incoming data. The digital synchronisation pattern in a preamble field of a transmission burst can be transferred directly into analogue DC levels. Then, the signal is fed to the phase shifter, a voltage controlled delay element, which corrects the data phase continuously. The adjusted phase is held until the next ATM cell with a synchronised pattern arrives. The circuit is designed to meet the constraint that each ATM cell is preceded by a synchronisation pattern which consists of three bits, i.e. 010.

The bit synchroniser consists mainly of integrators, comparators, track-and-hold circuits, and a phase shifter which were designed on the basis of advanced special RF circuit technologies. [9,15] Also, some auxiliary circuits are needed. In the following the basic principle of the bit synchroniser is described. Then, two main components of the synchroniser, the phase shifter and the anti-coincidence circuit are described in greater detail. Finally the performance of the circuit is discussed.

3.3.3.1 Basic principle

Figure 3.3.7 shows a principle block diagram of the bit synchroniser. The bit alignment function is achieved by using analogue techniques. Before the integration, both clock and data signals pass through a clipper to guarantee a nominal voltage level. The input data signals and the clock signals are then integrated simultaneously. Both positive and negative transitions of the preamble are converted into two ramps which are respectively compared with a ramp of the clock signal. The voltage differences obtained through comparators are sampled at the middle time point of the pulse cycle by using track-and-hold circuits. This time point corresponds to the time when 50% of the nominal voltage is reached. Then, the voltage difference is used as a control voltage for a VCPS (voltage controlled phase shifter) which delays the data signal according to the phase differences so that the signals are synchronised. The signal *SLOT* is used to reset the value of the integrated preamble after a transmission burst.

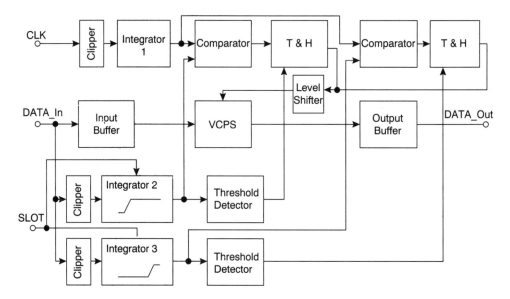

Figure 3.3.7
Principle block diagram of the bit synchroniser

3.3.3.2   *Phase shifter*

For the upstream synchronisation, the phase of signals is continuously delayed in a simple manner (i.e. via an externally applied voltage) over a span of time of 0–1608 ps, corresponding to a full clock period $(0–2\pi)$. Also, the phase shifter is designed to be integrated on a single chip using the same bipolar technology. Since a phase shifter circuit like the one shown in Figure 3.3.8 covers only a phase-shifting range from 0 to $-\pi/2$, it is essential to develop a concept for a circuit which can shift the phase over the full period continuously.

The basic principle for reaching such goal is to let the phase shifting be performed at half of the input frequency f, so that shifting the signal with frequency f by $-\pi$ can be reached by shift of a signal with halved frequency by only $-\pi/2$. The circuit shown in Figure 3.3.8 requires a signal with a frequency-independent phase difference of ϕ_{max} (here taking $-\pi/2$) with respect to the phase of the input signal. Together with the frequency division, this $-\pi/2$ signal can be generated in a simple manner. After phase shifting, the frequency must be doubled to get the desired value.

The different functional blocks of the phase shifter are shown in Figure 3.3.9. Frequency halving is performed by using a static frequency divider FH. The frequency divider also generates the frequency-independent phase difference of $-\pi/2$ required for phase shifting.

After generation of signals with the required phase difference by using a static frequency divider, phase shifting can be performed by

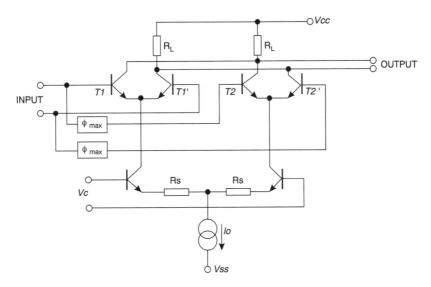

Figure 3.3.8
Phase shifter unit PS

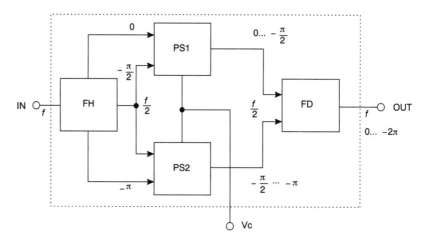

Figure 3.3.9
Block diagram of the phase shifter

applying the phase shifter unit PS shown in Figure 3.3.8. If the control voltage V_C has a sufficiently large positive value, the total current I_O flows through the transistor pair $T1$, $T1'$ and the output signal is only delayed with respect to the input signal by the electronic circuit delay. In this case the phase ϕ of the output signal is assumed to be zero. The maximum phase shift ϕ_{max} of the output signal, given by the time delay on the input, is reached for large negative values of V_C. Consequently, I_O flows completely through the transistor pair $T2$, $T2'$. Therefore, between the two extreme values, ϕ can be varied continuously by changing V_c.

The relationship between control voltage and phase of the output signal is given by the following formula:

$$V_C = V_t \ln \left(\frac{\tan \phi}{\sin \phi_{max} - \cos \phi_{max} \tan \phi} \right) + r I_O \left(\frac{\tan \phi / \tan(\phi_{max}/2) - 1}{\tan \phi \tan(\phi_{max}/2) + 1} \right)$$

with

$$r = R_s + r_e + r_b/\beta_0,$$

where V_t is the thermal voltage ($= kT/q$), r_e and r_b are the emitter and base resistances of the transistors, respectively, and β_0 is the current gain.

To extend the adjustable phase range from 0 to $-\pi$, two phase shifter units, PS1 and PS2 are used, respectively. The block FD (frequency doubling) performs the frequency doubling after phase shifting. The input signals of the FD are shifted in phase against each other by $\pi/2$ with respect to $f/2$. Therefore, the input and output signals of PS2 have to be shifted against those of PS1 by $-\pi/2$. This is achieved by using the corresponding signals from FH. Both phase shifter units are controlled by the voltage V_c. The FD drives an output signal with the original frequency f. The signals in the total phase range from 0 to -2π can be adjusted.

Since the control voltage V_c has a defined change range, the voltages sampled from track-and-hold circuits have to be converted into a voltage level that is suitable for the V_c. This is performed by using a level shifter.

3.3.3.3   Anti-coincidence circuit

Over the full clock period there is a critical time point when the sampling point of the data signal just coincides with the falling edge of the clock signal. This may lead to an undefined state of the circuit. The bit synchroniser contains an anti-coincidence circuit which can avoid the coincidence of the edges. In the following the principle is briefly described.

The circuit consists of peak generators, a peak comparator and a phase inverter. The waveforms are shown in Figure 3.3.10. Sampling becomes critical when the sampling point in the data signal (b) just coincides with the negative edge of the clock ramp (a). However, when this happens, the transition of the clock signal (d) and the transition of the data sampling signal (c) at the sampling time point become coincident. As a result, the peak comparator generates a toggle signal (e). The phase inverter changes the clock signal into its opposite phase (f), removing the conflict and allowing data to be clocked correctly. Moreover, the phase of the output data becomes aligned relative to the output clock.

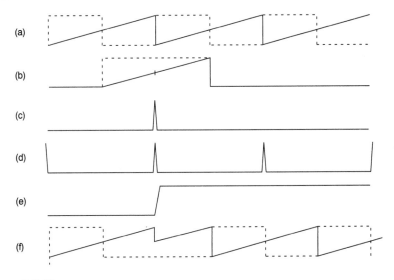

Figure 3.3.10
Timing diagram when using the anti-coincidence circuit (see text)

Figure 3.3.11
Reducing the noise through an integration

3.3.3.4 *Characteristics of the bit synchroniser*

To simulate the dynamic operation in a simple manner, the circuit was driven by a 010110... input data pulse train. To test the synchronisation behaviour, the input pulse trains were shifted in phase with respect to the clock. This input phase shift was simulated by using a delayed pulse source. Since the bit synchroniser operates in the full synchronisation range (2π), the output data stream and the clock signal are always in phase.

In the following, some characteristics of the bit synchroniser are described.

(1) Reducing noise   One of the main advantages of using an integrator is its ability to suppress undesirable spikes and fluctuations caused by external interferences, power supply ripple leaking in through the signal source, and by bipolar transistor flicker noise. Such spikes could lead to erroneous toggling when conventional digital circuits are used. In Figure 3.3.11 it can clearly be noticed that the different noise levels can be decreased to an insignificant value through an integration process. The bit synchroniser has a spike suppression feature that tolerates input data with noise spike voltages of up to 150% of the data signal level.

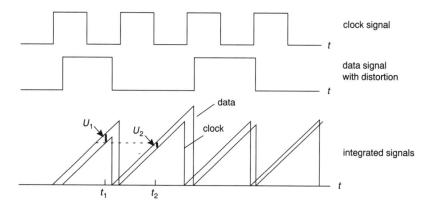

Figure 3.3.12
Two sampled points in two clock cycles

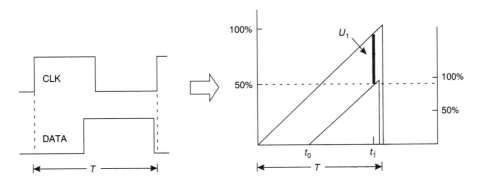

Figure 3.3.13
PWD worst case calculation

(2) **Tolerating pulse width distortion**　By sampling two voltages within two successive clock cycles, the bit synchroniser tolerates a large range of pulse width distortion (PWD). As Figure 3.3.12 shows, the voltages are sampled and held at the times t_1 and t_2 ($t_2 - t_1 = T$), respectively. After the voltages U_1 and U_2 are sampled and held, respectively, the mean of both voltages, namely $U = (U_1 + U_2)/2$ is then applied to the input of the phase shifter, so that a voltage/phase conversion can be made efficiently.

For the sake of reduction of noise, the sampling point should be set as late as possible in a clock cycle. However, if the waveform becomes very narrow due to PWD, the voltage of the ramp cannot be integrated to reach a higher level, i.e. a tolerance of such an extremely large distortion cannot be reached. Therefore, it is appropriate to choose a sampling point at the middle of a cycle.

The PWD worst case calculation can be carried out as shown in Figure 3.3.13 in which a 622 MHz operating frequency is assumed. The

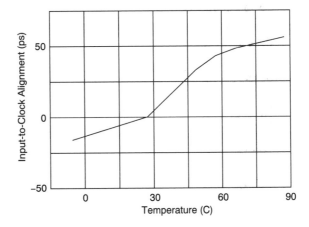

Figure 3.3.14
Timing phase shift vs. temperature

maximal PWD for the input signal to the bit synchroniser chip is

$$\text{PWD}_{\max} = t_1 - t_0 = T/2 = 804 \text{ ps.}$$

(3) Timing phase shift vs. temperature   An analysis of the influence of temperature changes shows that data alignment is well performed at 622 Mbit/s over a range from -5 to $+70°$ C (Figure 3.3.14). The result shows a maximal timing phase shift of $+11°$ if the phase difference between the incoming data and the clock signal is set to be $0°$ at 27°C temperature as the initial timing condition.

REFERENCES

[1]   Schrempp W and Sekimoto T, *Unique Word Detection in Digital Burst Communications*, Trans Comm Technol, **16**, 597–605, 1968

[2]   The work cited here was jointly performed by S Topliss, previously with Ascom Tech, Berne, Switzerland, now with GPT Ltd., PO Box 53, Coventry, CV3 1HJ, UK, D Beeler, with Ascom Tech, CH-3634, Hombrechtikon, Switzerland and L Altwegg, with Ascom Tech, CH-3018 Berne, Switzerland, private communication

[3]   Gardener F M, *Hangup in phase lock loops*, Trans Comm, **25**, 1210–14, 1977

[4]   Ota Y, Swartz R G, Archer V D, *et al*, *High-speed, burst mode, packet-capable optical receiver and instantaneous clock recovery for optical bus operation*, J Lightwave Technol, **12**, 325–31, 1994

[5]   Banu M and Dunlop A E, *Clock recovery circuits with instantaneous locking*, Electron Lett, **28**, 2127–30, 1992

[6]   Eldering C A, Herrerias-Martin F, Martin-Gomez R and Garcia-Arribas P J, *Digital burst mode clock recovery technique for fibre-optic systems*, J Lightwave Technol, **12**, 271–9, 1994

[7]   Dover R W and Pearson T, *Metastability and the ECLinPS™ Family*, Motorola Application Note AN1504, 1–8, 1991

[8]   Shear D, *Exorcise metastability from your design*, EDN, December 10, 58–64, 1992

[9]   Schmidt L and Rein H-M, *Continuously variable gigahertz phase-shifter IC covering more than one frequency decade*, IEEE J Solid-State Circuits, **27**, 854–62, 1992

[10] Gray P R and Meyer R G, *Analysis and Design of Analog Integrated Circuits*, John Wiley: New York 1984

[11] Rein H-M, *Multi-gigabit-per-second silicon bipolar ICs for future optical-fiber transmission systems*, IEEE J Solid-State Circuits, **23**, 664–75, 1988

[12] Al-Alaoui M A, "*A novel approach to designing a noninverting integrator with built-in low frequency stability, high frequency compensation, and high Q*," IEEE Trans on Instrumentation and Measurement, **38**, 1116–21, 1989

[13] Wang K C, Asbeck P M *et al*, *Voltage comparators implemented with GaAs/(GaAl)As heterojunction bipolar transistors*, Electron Lett., **21**, 807–8, 1985

[14] Souders T M, Schoenwetter H K and Hetrick P S, *Characterisation of a sampling voltage tracker for measuring fast, repetitive signals*, IEEE Trans on Instrumentation and Measurement, **36**, 956–60, 1987

[15] Gruber J, *et al*, *Electronic circuits for high bit rate digital fiber optic communication systems*, IEEE Trans on Comm, **26**, 1088–98, 1987

3.4 BURST MODE COMMUNICATION

3.4.1 Introduction

The BAF system uses cell based transmission techniques over a passive optical network with tree and branch topology. Simple baseband intensity modulation and a time division multiple access algorithm are used. As explained in Chapters 3.2 and 3.3 the transmission in the upstream direction is of the burst mode type. This has serious consequences for the transmitter in the ONU and the receiver in the OLT; which cannot be constructed in the same way as those in the conventional systems used for the downstream direction.

3.4.2 The burst mode transmission problem

3.4.2.1 *Optical transmitters' background*

A main requirement for optical transmitters in ATM PONs is the high bit rate capability. Moreover a relatively high optical output power and a small pulse distortion to allow connections over significant distances are needed. These requirements can be met by using semiconductor laser diodes for electrical to optical conversion in transmitters for ATM PONs. Laser diodes are threshold devices, in which the light output is proportional to the incremental current above threshold. Figure 3.4.1 shows the light output versus forward current graph, for different temperatures, of a 2 mW peak power InGaAsP Fabry–Perot single mode laser diode for digital optical communication systems. A suitable electronic circuit for driving such a device must include both a threshold current generator (I_{th}) and a data current generator (I_d). In normal operation the shift of the light–current diagram due to temperature variation has to be compensated for by the threshold current generator. A similar shift due to device aging is also expected.

Laser diodes are usually operated above threshold, where the slope $\Delta P / I_d$ is an important factor related to the differential quantum efficiency, i.e. the ratio between the number of generated photons and the injected electrons. From Figure 3.4.1 it can be seen that this slope is slightly or even not influenced by temperature variation in present laser diodes, but a degradation due to device aging is possible. In order to stabilise the optical output power most of the laser diodes commercially available are internally equipped with a monitor photodiode, a cooler element (usually based on a Peltier cell) and a thermistor for temperature monitoring.

A conventional laser driver block diagram [1], for continuous transmission application, is shown in Figure 3.4.2. A portion of the emitted light is directly coupled to the monitor photodiode inside

Access to B-ISDN via Passive Optical Networks. Edited by U. Killat
© 1996 John Wiley & Sons Ltd

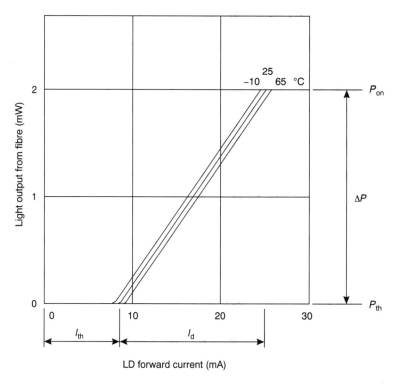

Figure 3.4.1
Light output vs. current graph

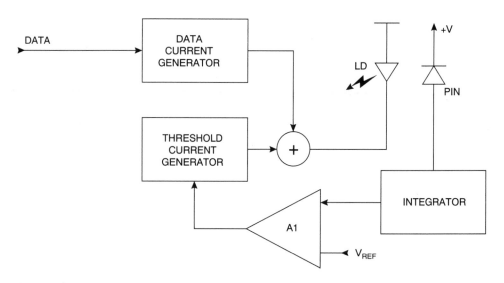

Figure 3.4.2
Single loopback continuous mode laser driver block diagram

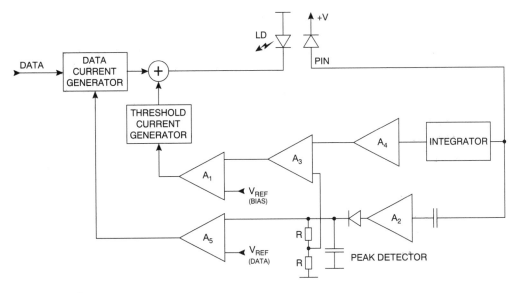

Figure 3.4.3
Double loopback continuous mode laser driver block diagram

the LD module. The integrated monitor signal, which is proportional to the mean value of the optical output power, is compared with a reference voltage. The error amplifier output controls the threshold current generator. The feedback loop bandwidth is usually very small to ensure good stability. Transmitted data need to be scrambled to assure a 50% duty cycle, and the data current generator is adjustable, but not controlled by any optical power information. A possible evolution of this laser driver (see Figure 3.4.3) can include a fast AC coupled monitor amplifier and peak detector [1], whose output signal is proportional to the peak data information, in a feedback loop controlling the data current generator.

As the amplifier A_2 is AC coupled one half of the peak level corresponds to the midpoint of the data modulation. This value is subtracted from the mean value of the output power obtaining a voltage level proportional to the threshold light, which is compared with a bias reference. In this case a possible differential quantum efficiency degradation can also be compensated for. Even in the double-loop configuration this kind of laser driver is just suitable for continuous transmission of scrambled data.

3.4.2.2   *Burst mode transmitters' requirements*

In a TDMA PON the ONU optical transmitter (subscriber side) has to cope with burst mode transmission, because the optical connection to the OLT is time-shared between the subscribers. As a first consequence the ONU transmitter optical output power control cannot be based on a mean value optical power control circuit. Indeed in the absence of data to be transmitted, a conventional laser driver, like the ones described before, would lead to a bias current of $I = I_{th} + 0.5I_d$. A different approach for the optical power control in burst mode transmitters (BMTs) is then necessary.

At the OLT side of a TDMA PON the incoming data packets from different ONUs are detected and amplified by a burst mode receiver (BMR). In some PON based communication systems the optical output level of the ONU transmitters is set to an appropriate value, depending on the branch attenuation, so that the packets' amplitude at the BMR input is always the same. In other systems the transmitted power is fixed, and the amplitude difference of the incoming packets has to be compensated for by the BMR in a defined range; this is the case for the BAF system. In the case of a transmitting ONU at the maximum distance allowed, the S/N ratio at the BMR input can be considerably worsened by the background light generated by closer ONU in the off condition ($P_{th} = -14 \div -18$ dBm). Moreover, in the BAF system one of the purposes of the guard time between packets is to allow the OLT receiver to sample a low-level signal, transmitted by an ONU during the ranging procedure (see Chapter 3.5). These considerations lead to the requirement of having all the ONU transmitters completely switched off between transmitted packets. In other words, it is not allowed to bring lasers to threshold before transmission of the first bit of a data packet.

In a BMT the laser diode has then to be modulated, at least at the beginning of a data packet, from $I = 0$ to $I = I_{on}$. When a laser is driven in such a way the optical emission is delayed by a quantity called the turn-on delay. In fact when the bias current is below threshold [2] the carrier density is very small, as is the number of emitted photons. The turn-on delay t_s is proportional to the carrier lifetime τ_e:

$$t_s = \tau_e \ln \left[\frac{(I_{on} - I_{off})}{(I_{on} - I_{th})} \right], \quad I_{off} < I_{th} < I_{on}$$

In laser diodes for optical communications τ_e is about 1.5 ns. In the case of $I_{off} = 0$, for the laser characterised in Figure 3.4.1, the estimated value of t_s is 580 ps. The turn-on delay value represents itself a limit for the transmission bit rate in burst mode with zero guard time between packets.

3.4.2.3 Turn-on delay compensation

Two compensation methods, both based on a pulse pre-distortion technique, will be considered. In the first case the laser diode is modulated

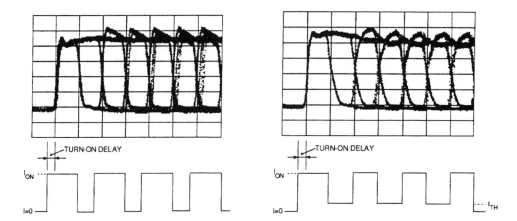

Figure 3.4.4
(a) zero bias modulation; (b) first bit predistortion

from $I = 0$ to $I = I_{on}$ for all the "1" bits of the packet, which are time-extended. The electrical pulse width at the laser input is equal to the bit period plus the turn-on delay (Figure 3.4.4a). In the second case (Figure 3.4.4b) the laser is modulated from $I = 0$ to $I = I_{on}$ for the first bit of the packet, which is extended, and from $I = I_{th}$ to $I = I_{on}$ for the following data bits. The threshold current is switched off at the falling edge of the last bit of the packet.

The first technique, later called zero bias modulation, can be achieved by a very simple electronic circuit but can be applied only if the laser turn-on delay can be considered as a constant. In reality this is not the case. When a laser is suddenly switched from below to above threshold, the time necessary to obtain a fixed output power varies in a random manner [3]. This effect is due to some broadening in the probability distribution of the number of photons in the laser cavity, caused by the spontaneous emission process during the transient following the switching. The result of this effect is a turn-on delay jitter, which is transformed into a time jitter on the rising edge of the optical pulses.

An influence of the data pattern [4] on turn-on delay jitter has also to be taken into account. For high bit rate transmission, if the electrical pulses are not separated by a sufficient amount of time, the laser cannot reach the off condition before the emission of the next pulse. Experiments on zero bias modulation performed with different lasers have shown very different values of the turn-on delay jitter, which is then dependent also on the device technology. Low-threshold laser diodes with a very high differential quantum efficiency, modulated with a continuous PRBS $2^{23} - 1$ zero bias data signal, have shown a huge 550 ps peak–peak turn-on delay jitter.

In the second technique described, later called first bit pre-distortion, only the first bit can be affected by turn-on delay jitter. A guard time is inserted between upstream packets, therefore the first bit of a packet occurs after a zeroes' sequence that is at least 8 bits long, which is the worst condition in the case of two consecutive packets being transmitted by the same ONU.

In Figure 3.4.4 different responses for the two compensation techniques described are shown, with particular reference to the initial bits of a packet. The test pattern is a PRBS $2^{23} - 1$ with an 8-bit guard time every 56 bytes and with the first bit following the guard time set to "1". The increase in time jitter due to the pattern effect is evident when the first technique is used. Moreover the guard time decreases the pattern effect on the first bit time jitter. These are the reasons why first bit pre-distortion has been implemented, eventhough more complicated electronics are required. Besides the input data signal, a cell envelope signal is needed by the laser driver (see Figure 3.4.5) in order to switch on the threshold current generator at the rising edge of the first bit of the packet and to turn it off at the falling edge of the last data bit.

3.4.2.4 BMT optical output power control

If a wide dynamic range BMR is used in a TDMA PON based system, a certain fluctuation of the ONU transmitters' output power can be tolerated. System reliability reasons lead to fixing a limit to this fluctuation. Moreover with the first bit pre-distortion technique the ONU laser is operated above threshold during the transmission of a packet, therefore the threshold current has to be set to an appropriate value to avoid pulse distortion or extinction ratio penalisation.

A −1 dB decrease of the ONU transmitter optical output power due to laser aging is allowed within ten years. The upstream cells' payload is not scrambled; in other words the mean value of the optical power during transmission is not a constant. Owing to the high upstream bit rate (622.08 Mbit/s) the design of a laser driver based on a loopback including a fast peak detector is a very difficult task. Although with commercial LD modules it is possible to achieve bit rates of more than 1 Gbit/s the built-in monitor photodiodes have in fact a limited bandwidth, usually below 100 MHz.

A first approach to the problem of optical power control in a BMT is based on LD temperature stabilisation. The driver shown in Figure 3.4.5 includes a turn-on delay compensation circuit, based on the first bit pre-distortion technique, and a temperature control circuit; no loopbacks on the data path are now considered.

The LD module temperature stabilisation prevents the light–current curve shifting due to temperature. The threshold current generator must be very stable in the operating temperature range, therefore the

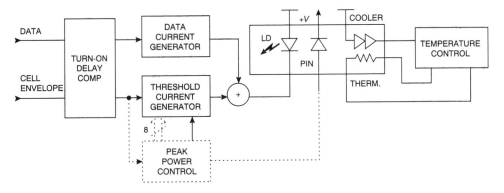

Figure 3.4.5
Burst mode transmitter block diagram

use of one-chip laser driver ICs is not recommended. The optical power variation due to device aging is not compensated for by using this BMT structure. In order to verify whether the BAF specification (−1 dB after 10 years) can be met just by keeping the temperature of a device like the one characterised in Figure 3.4.1 constant, the CNET RDF93 reliability model has been applied. The calculated MTTF (mean time to failure), defined as a 50% increase of the driving current for the same optical output power, for the considered device kept at 25 °C is 196.8 years.

From [5]:

$$\Delta I \propto t^{1/2}$$

a 10% driving current drift is expected after 7.95 years. The BAF requirement is not met just by keeping constant the LD temperature, so an additional peak power control has been developed.

Even if the LD monitor photodiode bandwidth is not wide enough to realise a fast peak detector, the transmitted peak optical power can be calculated considering a monitor current approximate step response like:

$$I_{\mathrm{mon}}(t) = I_{\mathrm{peak}} \cdot (1 - e^{-t/t})$$

The peak power control developed[†] is based on the fast acquisition of the value reached by the photodiode current $I_{\mathrm{mon}}(t)$ after the first three bits of the "111001011" unique word. This control finds application whenever an alignment word, with at least one initial "1" bit, is present in the system frame structure. The block diagram of such a control is shown in Figure 3.4.6.

The amplified response of the monitor photodiode is rectified and stored in a capacitor before the sampling. In this way a lower sampling instant accuracy is tolerated, obtaining a better measurement accuracy.

[†] Patent pending

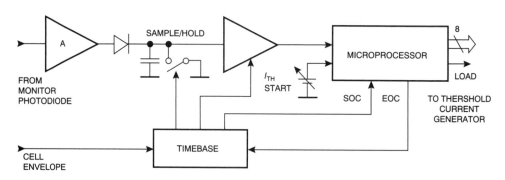

Figure 3.4.6
Peak power control block diagram

The held signal is sent to a single-chip microprocessor, which includes an A/D converter. The peak value is calculated and the obtained result is stored in an internal RAM. During the conversion, the acquisition of new samples is disabled by the EOC (end of conversion) signal. The mean value of the peak optical power is calculated over a number of samples set via software. If a correction is necessary, the new threshold current value is calculated and sent in an 8-bit format to the threshold current generator, which is activated only during the packet transmission by the cell envelope signal.

This signal also supplies the information about the beginning of the packets to a timebase, which generates suitable pulses for the hold, SOC (Start Of Conversion) and discharge capacitor functions. The initialisation procedure consists in loading the I_{th} start value, which corresponds to the laser start-of-life threshold current. The up-to-date current value will be set after the first calculation cycle. A possible upgrade could include a non-volatile memory to hold the last threshold current value even after a transmitter power switch-off.

The peak acquisition is not made on every packet, because of the microprocessor conversion time ($\approx$20 μs). This peak control finds then application when no sudden temperature variation is expected, or in combination with a temperature control circuit. Owing to the limited current drift expected this peak control has a single loop structure. In the case of differential quantum efficiency degradation a small extinction ratio decrease is expected. The major problem in realising a double loop configuration is again the development of a fast peak detector (see Figure 3.4.3), due to the limited monitor photodiode bandwidth.

3.4.2.5   External modulation BMT

In all the laser drivers described in the previous sections the data information is transformed in a current, which is directly injected into the laser diode (direct modulation). In the transmission technique described in Figure 3.4.7 the data information is sent to an optical

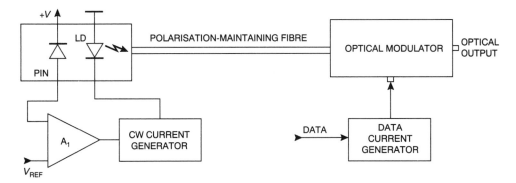

Figure 3.4.7
BMT with external modulator

modulator, which is put in series with a laser diode driven in continuous mode.

Present optical modulators are based on a solid state implementation of the Mach–Zehnder interferometer, using titanium diffused wave-guides on lithium niobate ($LiNbO_3$). Polarised light coupled into the input waveguide is divided at an input 3 dB splitter into two wave-guides. These run parallel to each other through two phase modulators, controlled by a common drive electrode. The voltage applied to this electrode causes the optical phase of one branch to be advanced and the other to be delayed, by the same amounts. These signals are then combined into a Y-junction, where in the case of equal phase they combine to launch the fundamental mode to the output waveguide. If the signals are out of phase they are transformed, in the junction, to a higher-order mode and lost into the substrate.

It is evident that in this transmitter there is no need of turn-on delay compensation, and the optical power emitted by the laser can easily be controlled via the monitor photodiode, which generates DC current information. On the other hand, optical modulators are sensitive to light polarisation, therefore the connection to the laser diode requires the use of polarisation maintaining fibre. Moreover the typical value of the extinction ratio is 20 dB; that means it is not possible to have zero emission between packets. The insertion loss is typically 4 dB; that implies the use of more powerful sources. The evolution of the transmitter shown in Figure 3.4.7 is represented by modulator inte-grated laser diode modules. These devices have been conceived for long transmission span (>100 km) and high bit rate (≈ 2.5 Gbit/s). A very low spectral width (<0.5 nm) allows operation in the 1550 nm region, where fibres show a very low attenuation, with a small penal-isation due to chromatic dispersion. The performance of the devices commercially available at the end of 1994 shows an extinction ratio of $10 \div 15$ dB, an optical output power of $+2 \div +6$ dBm.

For the time being the cost of optical modulators and modulator integrated LD modules is relatively high. Moreover, improvement of the extinction ratio and availability of 1300 nm devices are necessary for their application in TDMA PONs. In any case, the use of integrated optical modulators represents an attractive solution for the burst mode transmission problem.

3.4.3 The burst mode reception problem

In the upstream transmission path of the BAF system special techniques are required to deal with the bursty nature of this traffic. Owing to differences in optical attenuation in the different branches in the PON, cells arriving at the OLT can exhibit large power variations. The receiver at the OLT has to cope with these power variations and has to be able to receive the cells correctly, irrespective of their origin. In general, there are two ways to deal with these power variations.

- The optical power can be levelled at the transmitter side by adjusting the transmit power of the ONU transmitter, dependent on the optical attenuation in the path
- The use of a dedicated, optical burst mode receiver at the LT side, that is capable of handling short packets with large amplitude deviations.

From the cost point of view it is preferable to use as much standard equipment as possible in the ONU and deal with the optical burst mode problems on the OLT side. For the BAF system this approach has been taken.

Conventional lightwave receivers are AC coupled and therefore band limited. The DC contents of the received signal are not allowed to vary very much in time, because of the charging and decharging time of the capacitors in the various circuits. Change in the DC content of AC coupled receivers will result in bit errors due to an improper setting of the decision threshold in the decision circuitry. Scrambling or coding techniques are frequently used to decrease the low-frequency contents of the data signal.

When AC coupling is applied in the receiver, high sensitivities can be obtained. On the negative side, however, AC coupling significantly increases the complexity of the optical data link. Coding, which is required to shape the signal's spectrum, may take a fraction of the available bandwidth, but most importantly, AC coupling makes use of the optical data link inconvenient in certain applications. Especially with burst mode optical transmission as in the BAF system, there is a need for an optical link which appears to the user as an electrical link and, from the viewpoint of convenience, behaves as a galvanic connection. At the same time, one would like to retain the advantages of optical transmission, e.g. high transmission power budget, insensitivity to electrical (noise) disturbances, etc. In the BAF system with its stringent requirements, e.g. splitting factor of 32, maximum fibre length of 10 km, high transmission speed of 622 Mbit/s and the exacting receiver sensitivity, it is absolutely necessary to realise this special burst mode compatible optical link.

In Chapter 3.4.3.1 the principles of DC coupled burst mode reception will be discussed. This is based on the extensive work and analysis of R G Swartz [6]. The resulting implementation for the BAF system will also be presented. Applying AC coupling has certain advantages, as discussed above. In Chapter 3.4.2 an alternative burst mode receiver realised by K de Blok [7], based on AC coupling, will be described.

3.4.3.1   Burst mode receiver (DC coupled)

The obvious approach of just removing the AC coupling capacitors in a standard receiver to obtain a DC coupled version appears to be inadequate. Owing to the DC contents in the signal, saturation effects will occur in the amplifier chain. A better solution is, in addition to removing these capacitors, to amplify the *difference* between the data signal and a reference voltage. This reference voltage has to be centred exactly between the high and low values of the data signal (see Figure 3.4.8a). This should be done very precisely, to avoid excessive pulse width distortion (PWD) and, moreover, very quickly because the received optical power varies from cell to cell and is unknown in advance. So the required value of the reference voltage is dependent on the incoming optical power and needs to be continuously self-adjusting.

In Figure 3.4.8b both differential outputs of the preamplifier are used and fed to a differential input decision circuit. V_{ref} is generated by the feedback circuitry in the preamplifier. A new problem has now been introduced. To avoid instability at the outputs of the decision circuit when no light is being received, the V_{ref} must have an offset. This offset is called the logical zero offset (LZO).

In principle a small fixed offset in V_{ref} will suffice, dependent only on the noise levels at the inputs. To minimise pulse width distortion (PWD), the reference voltage value should be exactly one half of the output swing of the preamplifier. In this case logic zero and logic one pulse widths are equal, as shown in Figure 3.4.9a. When the input power increases and V_{ref} is not readjusted to one half of the output swing (Figure 3.4.9b) PWD will increase because the decision level is no longer centred in the middle of the pulses.

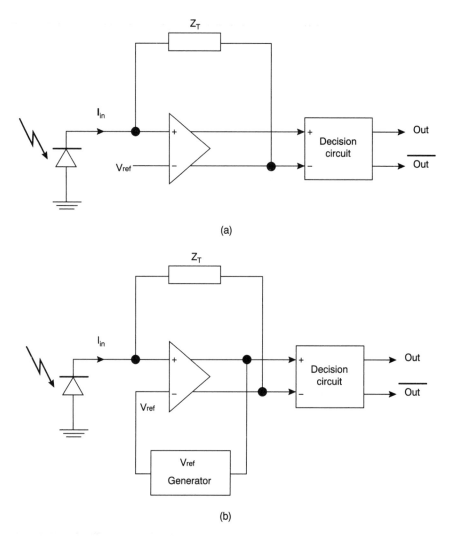

Figure 3.4.8
Preamplifier design (a) Fixed threshold; (b) Adaptive threshold

The basic circuit diagram of a differential preamplifier using the principle of readjustment of V_{ref} is depicted in Figure 3.4.10. The preamplifier is in fact built out of three blocks. The first is a differential transimpedance preamplifier A1 with feedback resistors Z_T. A1 should have a broad frequency response and very short propagation delays. The second block is a high-speed peak detector composed of differential amplifier A3, two transistors and a capacitor. The third circuit block is a post amplifier A2 to boost the output of the front-end receiver.

Sensitivity analysis   Sensitivity is defined as the minimum detectable optical power corresponding to a certain bit error rate, and is usually

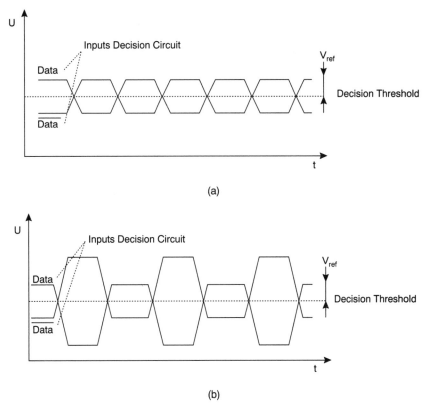

Figure 3.4.9
(a) Correct threshold, no PWD; (b) Incorrect threshold, introduction of PWD

determined by the noise characteristics of the receiver. The dominant noise source in AC coupled transimpedance receivers is the thermal noise of the feedback resistor Z_T. In this special DC coupled version two major noise sources can be added:

- the shot noise arising from the collector current of the two transistors in the peak detector circuit
- the shot noise from the base current of those two transistors.

Furthermore, the dynamic V_{ref} and LZO mechanism results in a theoretical sensitivity deterioration of 3 dB, as will be shown.

In Figure 3.4.11a the situation for a normal AC coupled receiver is sketched. The decision threshold is at the zero level and the input signal swings symmetrically around this level. The overall voltage swing is $I_{in} \cdot Z_T$.

In the case of burst mode reception, using the adaptive threshold setting, the preamplifier receiver output will behave as follows. With

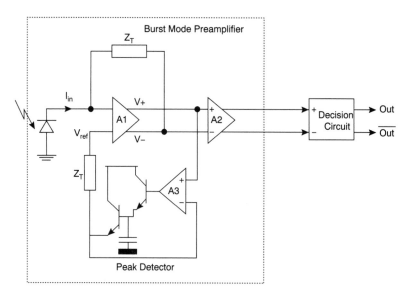

Figure 3.4.10
Basic circuit diagram of DC coupled receiver

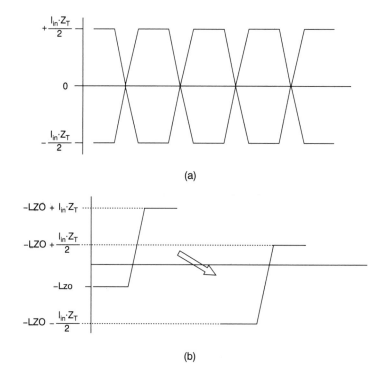

Figure 3.4.11
Signal and threshold levels in (a) normal AC coupled; (b) DC coupled burst mode
receivers

the arrival of the burst the first received "1" will rise instantaneously to an output level of $I_{in} \cdot Z_T - LZO$. However with acquisition of the pulse amplitude information this one level will gradually reduce to the level of $+1/2 \cdot I_{in} \cdot Z_T - LZO$. This phenomenon introduces a sensitivity penalty of exactly 3 dB.

Realisation   A dedicated multi chip DC coupled burst mode receiver (BMR) module has been developed in the BAF project, according to the principles discussed above. The BMR must be able to correctly receive consecutive cells that exhibit large differences in optical power. To keep the system flexible, the BMR operates without any knowledge of the optical power level of the next cell to be received. At the end of each cell reception, the decision level is reset; at the beginning of each cell the decision level is captured very quickly (within a time period of 2 bits). This requires a BMR module with a difficult, very-high-speed decision level adjustment and a large dynamic range.

Besides the correct reception of upstream data, the upstream coarse ranging tone must also be received properly. This additional requirement significantly increases the complexity of the BMR module. The upstream coarse ranging signal is a low-frequency tone which is superimposed on the data traffic. For data reception the ranging signal can be assumed to be constant, appearing as "DC"-background light. A dark current compensation (DCC) circuit has been implemented to remove the DC level in the data path. The photocurrent through the APD is sampled in the transmission gap (when no data is present) and compensated for during the reception of the next cell. To obtain the exact DCC value the realisation of a very fast and accurate sampling circuit is required. The DCC concept yields a receiver which is insensitive to low-frequency background light. The samples that are taken to compensate for the dark current are used to extract the ranging tone from the upstream data traffic.

In Figure 3.4.12 the block diagram of the BMR is depicted. The most complex parts are the sample-and-hold (S&H) circuit and the peak detector (dynamic reference voltage adjustment) circuit (PD). During the guard time the feedback input of the preamplifier is connected via the 2:1 multiplexer to a second reference voltage (V_{ref}) and the dark current is sampled. The sample is buffered to the coarse ranging output and the DCC circuit is adjusted. Thereafter, the 2:1 multiplexer switches to the decision level adjustment circuit. This feedback circuit consists of a peak detector that stores half of a peak, so the decision level is adjusted to half the peak amplitude of the data. The peak detector is discharged in each gap.

The module has been fully tested and found to work, in combination with the optical distribution network that is used in the BAF system demonstrator. The major problem encountered during testing

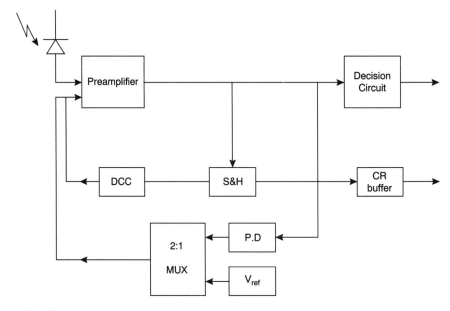

Figure 3.4.12
DC coupled burst mode receiver block diagram. PD, peak detector; CR, coarse ranging;
DCC, dark current compensation

was the temperature stability of the BMR. After the BMR has been
equipped with a Peltier cooler, its performance is stable over a temperature range of at least 45 °C. The sensitivity of the module is −25.5 dBm
over this temperature range. Within a temperature range of 10 °C a
sensitivity of −27.0 dBm has been reached. These results comply with
the requirements of the optical network that is going to be used in
the demonstrator, which has a maximum attenuation of 23 dB. The
minimum dynamic range of the BMR is 10.5 dB, so the maximum
optical input power can also be handled. The output eye pattern of
a cell with a repeated 16 bit pseudo-random pattern as a payload at
−26 dBm optical input power is shown in Figure 3.4.13.

The second function of the BMR is the extraction of the coarse
ranging tone. The ranging output has been measured with a specially
developed coarse ranging test generator. These measurements show a
coarse ranging output which fulfils the system requirements.

3.4.3.2   *Burst mode receiver (AC coupled)*

Design considerations   The DC component of the NRZ upstream
signal is highly dependent on the data and cell levels. This inhibits
the use of a fixed threshold for the reconstruction of data from the line
signal. In the DC coupled receiver an adaptive threshold is used to
handle this varying DC component. In this BMR, the DC component is

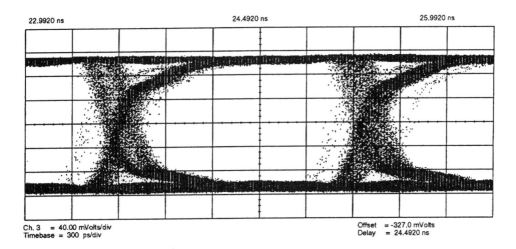

Ch. 3   = 40.00 mVolts/div
Timebase = 300 ps/div

Offset = -327.0 mVolts
Delay   = 24.4920 ns

Figure 3.4.13
Output eye pattern of DC coupled BMR

first removed before carrying out any level control or signal processing. The DC component is removed by differentiating the upstream NRZ signal. Differentiating results in a small loss of sensitivity. In practice, however, the effect of this loss is compensated for by the fact that differentiating makes the data path inherently insensitive to the low-frequency coarse ranging signal (10 kHz) and to temperature effects (AC coupling). Therefore no DCC is required, which simplifies the overall design. To avoid additional loss of sensitivity, level control is performed at the output of the low-noise preamplifier. Level control has to be very fast to adapt to the difference in cell amplitude within the gap time of a few bits. A feed-forward loop allows the response time of the control circuitry to be as fast as one bit, because the delay time in the loop can be fully compensated for by additional signal delay. A feed forward loop is therefore used in the AC coupled BMR. Detection of the coarse ranging signal and the upstream data signal processing are functionally separated to eliminate crosstalk between the two signal paths.

Implementation   In Figure 3.4.14 the basic configuration of the AC coupled BMR is shown.

Preamplifier   A low-noise preamplifier has been realised by using a second-order capacitive current–current feedback technique [7]. This has resulted in a wide dynamic range (20 dB$_{opt}$). The equivalent noise current of the preamplifier using a low cost PIN photodiode (1.2 pF) equals 2.2 pA/Hz. As a consequence of using a differentiator, the preamplifier may be AC coupled, which has the advantage of

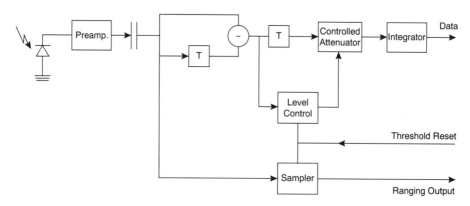

Figure 3.4.14
AC coupled burst mode receiver block diagram

eliminating thermal effects. The time constants of the (de)coupling capacitors have to be chosen much larger than the period of the coarse ranging signal (10 kHz) to avoid a phase offset.

Delay line differentiator   In this BMR, a delay line differentiator removes the DC component of the line signal before any signal processing or gain control is performed. Differentiation of the line signal introduces a small penalty in SNR, compared with ideal threshold detection. The signal U_s and the noise U_n from the preamplifier is split up into two paths, causing two mutually delayed signals with an amplitude of $(0.5U_s + 0.5U_n)$. The noise in both paths is not correlated due to this delay and adds on a power basis. The noise amplitude at the output is given by

$$\sqrt{2\left(\frac{1}{2}U_n\right)^2} = 0.707\ U_n$$

which results in a penalty of

$$10 \cdot \log\left(\frac{0.707}{0.5}\right) = 1.5\ \text{dB}$$

In addition, the differentiator acts as a high pass filter and blocks the low frequency coarse ranging signal from the data signal processing part.

Level control   The level control circuit is implemented by a voltage variable attenuator in a feed-forward control loop. To equalise the output of the BMR, the attenuation value is set according to the received optical power. The attenuator control signal is derived from the differentiated input signal by a fast peak detector. A part of the differentiated signal is fed via an emitter follower, to a Schottky diode and a small capacitor (10 pF). The capacitor will be charged to the peak value of the positive pulse. The charge of the capacitor is converted into a current (0–20 mA) proportional to the input level. This current controls the attenuator. Charging the capacitor takes less than 2 ns and the storage time is more than 200 ms. To reset the attenuator to its initial state, the capacitor is discharged within 2 ns by a shunt transistor. This shunt transistor is controlled by the threshold reset signal.

Set-reset integrator The (differentiated) output signal of the attenuator is fed to a set–reset integrator to restore the original NRZ data. Basically a set–reset integrator consists of two comparators, one with a positive threshold and one with a negative threshold. The output of the comparators controls a set–reset flip-flop. A positive pulse forces the flip-flop to the '1' state and a negative pulse to the '0' state. These functions are combined, in the receiver, into a single comparator with positive feedback.

Sampler To restore the low frequency ranging signal a sample of the line signal is taken during the gap between two cells. The sampler is actually a sample-and-hold circuit, implemented by a GaAs FET and a small capacitor. The sampler is controlled by the same signal used for the reset of the level control. The bandwidth of the sampler output signal ranges from 100 Hz to 30 MHz. The amplitude of the output signal ranges from 60 mV_{pp} up to 1.9 V_{pp} at coarse ranging levels of −40 dBm and −25 dBm respectively. The minimum SNR is 7 dB at an output bandwidth of about 30 MHz (at a data level of −15 dBm, ranging level −40 dBm).

REFERENCES

[1] Kressel H, Shumate PW and DiDomenico Jr M, *Semiconductor Devices for Optical Communication*, Springer Verlag, Berlin, Heidelberg, New York, 1980

[2] Petermann K, *Laser diode modulation and noise*, Kluwer Academic Publishers — KTK Scientific Publishers: Tokyo

[3] Spano P *et al*, *Experimental observation of time jitter in semiconductor laser turn-on*, Appl Phys Lett **52** (26), 1992

[4] Sapia A *et al*, *Pattern effects in time jitter of semiconductor lasers*, Appl. Phys. Lett. **61** (15), 1992

[5] Sim S P, *A Review of Reliability of III-V Opto-electronic Components, in Semiconductor Device Reliability*, NATO ASI Series E: Applied Sciences, **175**

[6] Swartz R G, *Electronics for high speed, burst mode optical communications*, Int J High Speed Electron, **1**, (3,4) 223–43, 1990

[7] De Blok C M, van den Brink R F M, van Vaalen M J M and Prinz P M J, *Fast low cost feed-forward burst mode optical receiver for 622 Mb/s*, EFOC&N '93 Proceedings, 256–60, 1993

3.5 THE RANGING PROBLEM

3.5.1 Introduction

The application of cell based transmission techniques over a passive optical network (PON) with a tree architecture, using intensity baseband modulation and regular time division multiplex techniques, will cause collisions in the network. Collisions occur when cells from different optical network units (ONUs), transmitted upstream, arrive at the same time, or partially overlapping in time, in the passive splitting/combining points of the optical network. These collisions are caused by the differences in optical path lengths in the different branches. These collided cells are propagated further upstream until they arrive at the optical line termination (OLT). The OLT cannot properly detect the bits of the collided cells, and a large number of bit errors and possibly cell loss will occur.

In PONs with a tree architecture, the ONUs are placed at different distances from the OLT. For the BAF system, these distances can vary between a maximum optical path length of 10 km and a minimum of 0 km. These distances result in a total round trip delay time of 0 to 0.1 ms. Downstream a continuous information flow, based on ATM-like BAF cells, is transmitted by the OLT and received by each ONU. Cells intended for a particular ONU are selected by means of address recognition by that ONU. For upstream transmission the described collisions have to be avoided. For that reason dedicated processing equipment has to be introduced. This equipment should be installed in the OLT because the OLT controls the flow of all upstream traffic. Procedures will be necessary to transfer control information between the different nodes in order to prevent upstream cell collision. A protocol to provide a properly functioning time division multiple access (TDMA) mechanism has to be developed for this purpose. With such a time division multiple access technique for upstream traffic, each ONU has to process the upstream data in such a way that optical signal collisions are prevented. This can be done by placing all ONUs at the same virtual distance from the OLT. The procedure of measuring the distance and placing all ONUs at the same virtual distance is called ranging.

In the BAF system, the total ranging procedure is split into three separate subprocedures. These are called coarse ranging, static fine ranging and dynamic fine ranging. The result of the total ranging procedure is that the cells arriving at the OLT are placed in their allocated slots within plus or minus one bit time. Ranging requires knowledge of the propagation delay between each ONU and OLT. In the BAF system, the delay measurement for the coarse ranging procedure is performed by means of a novel technique, which has not been applied to any

Access to B-ISDN via Passive Optical Networks. Edited by U. Killat
© 1996 John Wiley & Sons Ltd

other system. This novel technique, called out-of-band low-level, low-frequency ranging, is based on an analogue phase measurement of a low-frequency, low-level sine wave. This sine wave is transmitted as an intensity modulated light wave from ONU to OLT. After the measurement of the sine wave phase, a delay line is set in the ONU to be ranged in such a way that a certain predefined total delay from OLT to ONU and back is obtained. This predefined delay is determined by the maximum network length of 10 km. After the completion of the CR ranging procedure, the predefined delay is set within an accuracy of half a BAF-ATM cell time. The next ranging steps, static fine ranging and dynamic fine ranging, are based on measuring the arrival time of cells. This results in an accuracy of plus or minus one bit time when the total ranging procedure is completed.

Coarse ranging and static fine ranging are both so-called static ranging procedures, because ranging is performed in the absence of user traffic to the ONU being ranged. Static ranging is performed at initial system start-up, every time a new subscriber is connected to the network and, in case of an ONU failure, when an ONU restart has to be made. Dynamic ranging is performed continuously to overcome delay variations in the optical network and electronics, e.g. due to temperature changes of the fibre. This ranging procedure is active during regular cell transmission.

With a maximum round trip distance of 20 km, the total delay, expressed in bits, is maximally about 60 kbit at 622 Mbit/s. This is the number of bits which actually are on their way in the fibre, indicating also the losses in case of possible collision conditions. Therefore the ranging process has to be reliable and accurate. The ranging processing hardware and software, as implemented in the BAF system, behaves according to those requirements. Moreover the ranging circuitry is based on regular components.

In the following subsections a short description will be given of existing ranging methods, after which the BAF method will be described in more detail. The coarse ranging procedure will be covered in detail because of its novelty in communication systems. CR accuracy calculations and implementation issues will be described. Fine ranging, which is based on the arrival times of cells, is already being used in a number of PON systems and will be described somewhat less extensively.

3.5.2   Existing ranging methods

Generally, the ranging procedure starts without any knowledge of the distance to the ONU to be ranged. The first step will, therefore, be a coarse distance measurement. There are, in general, two methods available for performing the first step in the ranging procedure [4]. One method uses a silent period in all the upstream data traffic, and is called idle period ranging. A second method uses a low-level signal during the normal upstream data traffic of the ONUs which are already installed.

The use of idle period ranging is the most straightforward method. The normal upstream data traffic is stopped for all connections and a dedicated pulse travels from the ONU to the OLT. The delay time is measured and an appropriate delay line setting in the ONU under ranging can be installed. One consequence of this technique is that cells have to be buffered in the (remaining) ONUs for the round-trip time of 0.1 ms, so large cell buffer sizes are required. As ranging takes such a relatively long time, important traffic behaviour characteristics such as cell transit delay time and cell delay variation will show serious degradations. These degradations will be large enough to question whether normal ATM requirements concerning cell delay will be met. In the case of idle period ranging, 31 ONUs have to interrupt the upstream ATM traffic for 0.1 ms when an ONU is being ranged.

The low-level static coarse ranging method is based on spread-spectrum techniques. This technique measures the distance between ONU and OLT with the aid of a pseudo-random bit pattern. A correlation measurement system will give the distance. Using this ranging method, traffic from all the other ONUs can continue. Unfortunately, such correlation techniques require complicated electronic hardware and result in a rather low ranging accuracy. The next step in the ranging procedure, the static fine procedure, will still require a large idle time slot.

Considering the disadvantages of both the systems described above, a new ranging method has been developed for the BAF system.

3.5.3   Description of ranging procedures in the BAF system

3.5.3.1   Coarse ranging

This is the first step of the ranging process. The coarse ranging (CR) method is performed by a very accurate phase measurement at the OLT of a low-level, low-frequency sine wave signal. This signal is transmitted by the ONU that is to be ranged, the target ONU. The frequency of the ranging signal is determined by the distance to be covered. With a maximum distance between OLT and ONU of 10 km, the maximum round-trip time for a signal travelling from OLT to ONU and back can be calculated to be 0.1 ms. With this maximum round-trip time the frequency of the coarse ranging signal must be equal to or smaller than 10 kHz to maintain an unambiguous relationship between phase and distance.

This coarse ranging sine wave signal is transmitted by the target ONU when a downstream OAM CR command is received by the ONU. This command is transported as a bit in the downstream preamble, in the so-called fast OAM channel. In the ONU this 'start coarse ranging' command will trigger the CR signal generator, which is a digital processing generator, starting with a defined start phase. This signal is superimposed on the laser bias current with an amplitude much smaller than the data drive current. So the upstream optical signal, received by the OLT, is composed of a low frequency ranging signal and a high frequency data signal coming from normally operating ONUs. This composite signal is split into the originating signals at the OLT receiver. The coarse ranging signal will be sampled in the gap between the packets, thus avoiding disturbances due to the burst mode effects of the incoming cells. This process is depicted in Figure 3.5.1.

In this way, only the noise of the receiver affects the coarse ranging signal. The phase of the coarse ranging signal can be very accurately measured by digital processing in the OLT. The reference for the phase measurement is the start time when the OAM command is sent downstream. To obtain an undisturbed data signal, the CR signal level will be removed from the composite signal by dedicated circuits (see Chapter 3.4). The accuracy of the CR procedure allows the total network delay, when the delay line in the ONU is set, to be established within half a BAF-ATM cell time. The next steps in the ranging procedure provide a further refinement of the ranging.

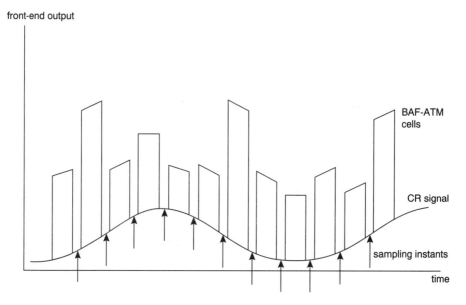

Figure 3.5.1
Coarse ranging sampling method

3.5.3.2   Static fine ranging

Static fine ranging (SFR) is the second step in the ranging procedure. The SFR process is based on digital measurement techniques. One cell slot is allocated to the target ONU for transmitting a special SFR cell. This is a small cell consisting of only the preamble with a length of 9 bits. This cell will be placed in the centre of the time slot allocated to the target ONU.

The SFR process consists of two subprocesses: the adjust phase and the verify phase. The adjust phase procedure is started by sending an OAM cell downstream to trigger the SFR actions. The OAM cell contains the delay line setting for the target ONU, obtained from the coarse ranging procedure. The ONU transmits the SFR cell after a permit from the OLT has been received. Once the SFR cell is detected by the OLT, its position relative to the allocated slot is measured by a digital counting technique. This position is determined with a precision of one bit. With this value, the delay line in the target ONU is adjusted in such a way that the total BAF network round-trip time equals the predefined round-trip time with an accuracy of one bit or 1.6 ns.

The next SFR step is the verify phase. This is the verification of the adjust phase process. Verification is a check on whether the SFR cell is indeed in the centre of the allocated cell slot. This is required because bit errors may occur in the transmission, resulting in an inaccurate SFR result. This error possibility is very small, but a false fine ranging value will lead to cell collisions from which it may be difficult to recover. When the verification procedure succeeds, the ONU can be considered accurately installed and normal cell based operation can proceed. Delay drifts of the optical path and of the electronic circuitry, due to temperature variations and aging, will be corrected by dynamic fine ranging.

3.5.3.3   Dynamic fine ranging

Dynamic fine ranging (DFR) is a continuous process to be performed on each cell entering the OLT. When static fine ranging is finished, the OLT will send permits for OAM cells to the target ONU. If the OAM cells are correctly received, normal traffic can start. To prevent undetected drifting of the round-trip delay, a certain minimum refresh rate is required. ONUs which transmit normal data traffic at a very low rate will be forced to transmit OAM cells. The DFR mechanism is based on the position of the preamble, also called the unique word (key word), at the beginning of the cell. This position is detected within a certain window, and deviations from the expected position lead to corrective action by dynamic fine ranging, which adjusts the delay in the ONU and recovers the normal cell position within its time slot. In this way the normal gap between two consecutive cells will be maintained. The actual length and values of the gap and preamble can be found in Chapter 3.3.

If the preamble detection of one cell fails, there exists the possibility of a cell collision in the optical network. This may possibly lead to more collisions, with the result that the data traffic of a number of ONUs is blocked. Therefore, failure of unique word detection must lead to a very fast switch-off of the ONU involved. This has to be performed autonomously by hardware via the fast downstream OAM channel.

3.5.4 Ranging accuracy considerations

This section deals specifically with the sources of inaccuracy in the coarse ranging procedure. The other ranging procedures, static fine ranging and dynamic fine ranging, are based on arrival times of cells and, owing to their digital characteristics, can be executed with a one-bit accuracy. Coarse ranging is based on the measurement of the phase delay of a sine wave and is, therefore, disturbed by such analogue effects as receiver noise. The CR accuracy requirement is calculated first. CR has to be sufficiently accurate that the next ranging step, dynamic fine ranging, can be applied within its measurement limits.

There are two error sources concerning CR: dynamic errors and static errors. Dynamic errors are caused by random processes, resulting in a certain statistical distribution of the CR result. Such errors are caused by thermal receiver noise, photodiode shot noise, quantisation noise of the analogue-to-digital converter (A/D converter), and data interference. Compensation for this type of error in the digital processing is in principal impossible due to the random nature of these effects. Static errors are essentially fixed errors due to the delay difference between the signal paths of the CR signal and the data signal. The main source of error is the phase shift in the CR sampling and amplification circuits. This occurs mainly because of bandwidth limitations on the low-frequency side, due to coupling capacitors. Another possible cause is the delay time introduced by the various digital processes involved in coarse ranging signal transmission and reception. Compensating for static errors using the processing software is possible because the errors are relatively invariant. This is, however, only applicable when the errors are reproducible for all the equipment to be made, and when the errors are not temperature or aging dependent. This is, in fact, only true for digitally introduced errors. Analogue phase errors must, therefore, be within the DFR required limits. Another possible error source can be a non-fixed relation between the CR frequency and the data bit frequency. To avoid this error, the CR frequency is derived from the data bit clock. Its actual value is the bit clock frequency of 622.08 MHz divided by 67200, or 9.25714 kHz.

The SFR procedure requires that the centre of a time slot for one cell must be determined with an accuracy of +/−0.4 BAF-ATM cell slot time of 720 ns. Expressed in time, this means an accuracy of +/−288 ns. With a CR frequency of 9.25714 kHz, the 288 ns time limit leads to a phase accuracy requirement of +/−17×10^{-3} radians, or about 1 degree.

This value has to be divided between static and dynamic errors. For the BAF design, it was decided that the static errors should be very small compared with the dynamic errors, leaving nearly the whole error margin available for dynamic errors. This requirement imposes heavy, but realisable, design consequences on the CR circuits.

Dynamic errors are assumed to have the normal or Gaussian distribution because they are, for the main part, caused by thermal receiver noise, as will be shown later.

Having a ranging power carrier level of S, embedded in Gaussian distributed noise with a power level of N, the standard deviation of the phase error σ_e occurring on the measured CR phase measurement is [1]:

$$\sigma_e = \sqrt{\frac{N}{S}} \tag{1}$$

Owing to this Gaussian distribution, a certain probability has to be accepted that the required accuracy is not obtained. This probability is set at 1%, so the chance of success is 99%.

Having a ranging result r, and a probability function $P(r)$, the following equation must be fulfilled:

$$P(r < \mu - k\sigma_e) = P(r < \mu + k\sigma_e) = 0.005 \tag{2}$$

This equation is valid for $k = 2.6$, so $2.6 \times \sigma_e$ equals 288 ns. This results in a CR standard deviation of 110 ns, which agrees with 6.6×10^{-3} radians or about 0.4 degrees at the ranging frequency. For the 1% that exceeds the 288 ns accuracy, the DFR fails, and the CR procedure must be performed again.

In Figure 3.5.2, the probability distribution function of the CR measurement result is depicted. The mean value of the CR result is μ, which agrees with the mean CR signal delay time. The maximum allowable error is +/−288 ns, the 1% area shows the probability of exceeding the maximum. The actual CR standard deviation after the optical/electrical conversion and amplification is much larger than the required value, as will be shown later. Further filtering is therefore necessary. This filtering can only be performed digitally because analogue filtering cannot obtain the required static accuracy.

Signal to noise ratio (SNR, S/N) and filter requirements are mostly expressed in dB. Having a requirement of $\sigma_e = 6.6 \times 10^{-3}$ rad the

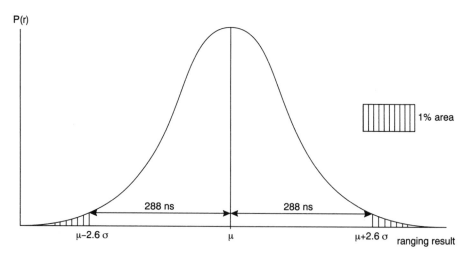

Figure 3.5.2
Probability distribution function of coarse ranging result

required SNR in dB can be calculated to be:

$$\frac{S}{N} = -20\log_{10}(6.6 \times 10^{-3}) = 44 \text{ dB} \tag{3}$$

The required SNR is 44 dB, while the obtained SNR after the O/E conversion and amplification is about 7 dB (Section 3.5.5.2). So the digital filtering has to improve the SNR by at least 37 dB.

3.5.5 Implementation of coarse ranging in the BAF system

3.5.5.1 *Ranging signal generation in the ONU*

Coarse ranging is based on an absolute phase measurement, therefore the highest care must be taken to prevent an additional phase shift due to hardware imperfections or limitations. Amplifier bandwidth, coupling capacitors and the temperature variations of such components can all create a phase shift. This static phase invariance is a major requirement for both OLT and ONU. In the ONU, it is the only source of inaccuracy, and the requirement is that this static error is negligible relative to the overall accuracy requirement.

The optical transmitter in the ONU has been designed to transmit either BAF cells or the ranging signal. Prior to transmitting the ranging signal, there is a ranging preset OAM command, coming from the OLT, to set the laser output at a ranging bias level of 35 μW. This value has been chosen because it is the threshold value of the applied laser type. Choosing a lower level will lead to a sharp increase in distortion on the lower side of the sine wave. An internal ramp generator is applied to perform this setting gradually, because fast changes of the upstream optical power can cause bit errors in the reception of data traffic from the other ONUs.

After the laser has been set to the ranging bias level, the low-frequency signal is transmitted as soon as the start CR OAM command is received. The sine wave, which is digitally generated, starts at a phase of −90 degrees, and is superimposed on the bias signal. In this way the optical ranging signal starts from its lowest level, which is the bias level, and reaches a peak value of 220 μW, without sudden steps in the optical level.

The sine wave is automatically switched off by the ONU after a long enough time is provided to measure the phase accurately. The switch-off procedure is also done gradually. The total transmitted coarse ranging signal is depicted in Figure 3.5.3. The mean value of the ranging signal transmitted by the ONU is about 110 μW or −10 dBm. Due to the fact that the laser is applied just above its threshold, this signal is somewhat distorted. This distortion has been measured, after optical to electrical conversion, by means of a spectrum analyser. A number of higher harmonic signals is present. The second harmonic has the highest level of about −30 dB relative to the level of the fundamental frequency. The levels of the higher harmonic signals decrease slowly with the harmonic number.

The distortion must not be allowed to influence the coarse ranging accuracy. Therefore, the digital processing in the receiver has to be designed in such a way that it is not sensitive to harmonic distortion. A

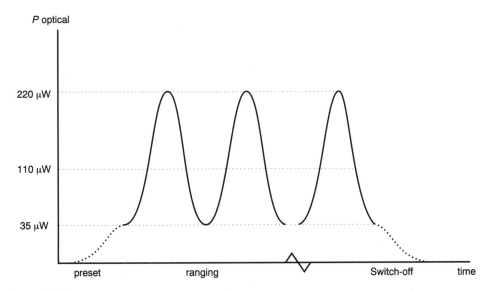

Figure 3.5.3
Optical NU output during coarse ranging

suitable algorithm is implemented which is robust enough to calculate a reliable ranging result if there is not too much hard limiting of the sine wave ranging signal. In that case, inaccuracies in the measurement of the ranging signal will be minimal. Otherwise, the clipping will cause a considerable decrease in the signal amplitude of the CR fundamental frequency resulting in an increase of the inaccuracy. It is clear that the bias level of the laser has to be controlled in order to maintain a stable optical level. Separate control systems for the ranging and the data bias level have been implemented.

3.5.5.2 *Ranging signal reception in the OLT*

Optical-to-electrical conversion The coarse ranging signal level at the receiver is dependent on the optical loss in the BAF network. This loss has a value of between 15 dB and 30 dB. The loss range is calculated by the addition of the best- and worst-case loss values of the optical splitters applied in the network, including their variability, and the fibre loss variability due to the distance variation from 0 to 10 km. As a consequence of the values given above, the ranging signal level at the OLT can have any value between −25 dBm and −40 dBm.

In the OLT, the reception of the ranging signal is performed by the burst mode receiver (BMR). This type of receiver is capable of receiving cells with a large variation in amplitude, even when those cells are superimposed on a the low-frequency ranging signal. The BMR also converts the ranging signal from an optical to an electrical signal. Mutual crosstalk between data and ranging signals is very small and does not cause any system deterioration. Two types of BMR can be implemented: one based on a PIN-FET combination and one based on a combination of an APD and silicon IC. Both amplifiers have approximately the same sensitivity for data and ranging signals.

The signal to noise ratio of the received ranging signal at the output of the receiver amplifier, $(S/N)_{\text{ran}}$, will be calculated for both receiver types being used in the BAF system. As is usual for optical receivers, this calculation is based on the so-called equivalent input noise model, which means that all noise sources are transferred to the input of the amplifier, at the interface between photodiode and amplifier. In other words, noise sources are, therefore, collected to that interface, and the SNR at the input equals now the output SNR because the amplification can be considered noise free. The calculation is based on the photodiode and amplifier performance given in Table 3.5.1.

For the PIN diode, parameters M and F are not applicable and can be set at a value of 1 for calculation purposes. The APD excess noise is dependent on M, the value for F in Table 3.5.1 is valid at $M = 10$.

The calculation will be based on the lowest coarse ranging level at the receiver input, because at this level the lowest SNR occurs. The

Table 3.5.1
Optical receiver properties

Receiver parameter	APD Si IC receiver	PIN-FET receiver
APD/PIN responsivity R	0.9 A/W	0.9 A/W
APD multiplication factor M	10	not applicable
APD excess noise factor F	4.5	not applicable
Amplifier noise density $\sqrt{\langle i_{eq}^2 \rangle}$	15 pA/$\sqrt{\text{Hz}}$	2 pA/$\sqrt{\text{Hz}}$
Amplifier bandwidth B	400 MHz	400 MHz

noise types taken into account are thermal receiver noise and photo-diode shot noise. In the BAF system all ONU lasers are switched off in the gap, therefore the photodiode shot noise is only determined by the received coarse ranging signal, and calculation will show that shot noise is almost negligible.

Both noise sources are assumed to have a flat spectral density behaviour over the frequency band of interest. Noise amplitude is assumed to have a Gaussian distribution statistical behaviour. Noise and signal currents will be calculated in time-averaged mean-square values: $\langle i^2 \rangle$. Noise sources are uncorrelated and mean-square values can be added. The amplifier input noise density, given in Table 3.5.1, is defined as a root-mean-square value.

The equivalent input mean-square thermal noise current $\langle i_{th}^2 \rangle$ of the amplifier is:

$$\langle i_{th}^2 \rangle = \langle i_{eq}^2 \rangle B \qquad (4)$$

Shot noise in photodiodes is caused by the primary current level in the diode. For the BAF system this current is only caused by the target ONU. All other lasers are switched off during the gap. This results in a mean ranging level P_{ran} of −40 dBm or 10^{-7} W,

The photodiode mean-square shot noise $\langle i_{shot}^2 \rangle$ is:

$$\langle i_{shot}^2 \rangle = 2eRP_{ran}M^2FB \qquad (5)$$

The electron charge e is 1.6×10^{-19} C.

For calculation of the ranging signal current at the receiver input, the mean-square of the optical CR sine wave has to be taken into account, which is approximately $P_{ran}/\sqrt{2}$.

$$\langle i_{ran}^2 \rangle = (RMP_{ran}/\sqrt{2})^2 \qquad (6)$$

The signal-to-noise ratio of the ranging signal at the receiver output is:

$$\left(\frac{S}{N}\right)_{ran} = 10\log_{10}\left(\frac{\langle i_{ran}^2 \rangle}{\langle i_{th}^2 \rangle + \langle i_{shot}^2 \rangle}\right) \text{ dB} \qquad (7)$$

Table 3.5.2
Signal and noise behaviour of BAF optical burst mode receivers

Receiver parameter	APD Si IC receiver	PIN-FET receiver
amplifier thermal noise $\langle i_{th}^2 \rangle$	9×10^{-14} A^2	1.6×10^{-15} A^2
photodiode shot noise $\langle i_{shot}^2 \rangle$	5.2×10^{-15} A^2	1.2×10^{-17} A^2
ranging signal $\langle i_{ran}^2 \rangle$	3.9×10^{-13} A^2	3.9×10^{-15} A^2
signal to noise ratio $(S/N)_{ran}$	6.2 dB	3.9 dB

The results of these calculations are given in the Table 3.5.2.

Measurements on the ranging SNR have been performed on a number of BMRs of both types. SNRs varying between 5 dB and 10 dB have been measured, with a mean value of 7 dB. No significant differences between the two types of BMR have been observed. The fact that the measured SNR is better than the calculated SNR in almost all cases is probably due to a small filtering effect at the input of the sampler. The calculated as well as the measured SNR indicate that the realisable SNR is much smaller than the required SNR of 44 dB (Chapter 3.5.4). Further filtering is thus necessary, and will be done by digital processing.

Sample-and-hold circuit   After optical/electrical conversion and some amplification in the receiver front-end, the signal has to be split into a ranging signal and data signal. The data processing part, describing the burst mode reception problem, is discussion in Chapter 3.4. The ranging signal is obtained by sampling the front-end output in the gap between the cells. This gap, or guard time, is normally 8 bits long: about 13 ns. Just in the middle of this gap, a sample of about 3 ns length is taken. A fast sample-and-hold switch takes care of this action. The fabrication technology of this switch depends on the BMR type, it is either a silicon IC, or a discrete high-frequency gallium–arsenide field effect transistor. The timing of the sampling action is depicted in Figure 3.5.1.

The sampling rate is equal to the BAF cell rate of 1.39 MHz. This sampling frequency is fast enough to reconstruct a coarse ranging signal of about 9 kHz. This ranging signal is embedded in noise with a bandwidth of 400 MHz. Filtering the O/E converter output signal, before sampling, will spread out the data signal into the gap, creating crosstalk between the ranging signal and the data. This effect decreases the coarse ranging accuracy and, because of the unknown statistical behaviour of the BAF cells, the disturbance effect cannot be calculated. If the O/E converter output is not filtered before sampling, aliasing occurs, and only the noise signal will be folded into the 1.39 MHz bandwidth. This effect can be calculated rather accurately because the noise level is known. For the reasons mentioned above the BAF system does not apply filtering of the signal prior to sampling. Owing to the digital processing applied, the aliasing effect can be dealt with and the required accuracy can be obtained.

For the calculation of the aliasing effect on the noise, the noise signal shape before and after the sample-and-hold action has to be considered. Before sampling, the noise has a mean-square value of $\langle i_n^2 \rangle = \langle i_{th}^2 \rangle + \langle i_{shot}^2 \rangle$, with a bandwidth $B = 400$ MHz and a noise spectral density of $\langle i_n^2 \rangle / B$. Note that these levels are defined at the interface between photodiode and receiver amplifier, which is also the reference interface for all further calculations. After the sampling-and-hold action the signal has the shape of a random digital wave, which is an ensemble of digital pulses having the same mean-square value of $\langle i_n^2 \rangle$ as before sampling, and a pulse frequency equal to the sample frequency f_s.

The spectral density $S(f)$ has the following shape [2]:

$$S(f) = \frac{\langle i_n^2 \rangle}{f_s} \operatorname{sin} c^2 \left(\frac{f}{f_s} \right) \qquad (8)$$

The spectral density has a first zero at the sampling frequency f_s of 1.39 MHz, which value can be considered as the approximate bandwidth of the noise signal after sampling. At a CR frequency of 9 kHz the sinc function is about 1, and the sampler output noise density is given by the first part of (8): $\langle i_n^2 \rangle / f_s$. The sampler input noise density is $\langle i_n^2 \rangle / B$. The aliasing effect causes an increase of the noise spectral density of B/f_s or the ratio of the receiver bandwidth to sampling frequency. This is a factor of 400 MHz/1.39 MHz or 288. The SNR of 7 dB is not changed during the sampling action because neither the signal nor the noise level is changed. The increase in spectral density is cancelled by the decrease of the noise signal bandwidth.

Analogue-to-digital conversion   After the sample-and-hold action, the signal is converted from analogue to digital samples by a high-speed 12-bit analogue to digital converter (A/D converter). These digital samples are stored as 12-bit words in a random access memory. When the RAM has been filled with enough samples, the on-board microprocessor will perform the required processing. The A/D converter process adds quantisation noise, which is dependent on the resolution of the A/D converter. Using a 12-bit A/D converter, the number of quantisation levels equals $q = 2^{12}$.

The maximum variation in network loss is 15 dB optical. This is equivalent to a voltage level variation, after O/E conversion, of a factor of 32. The maximum ranging signal will use the full A/D converter range. The minimum signal level is a factor of 32 smaller, and gives a relative power level S of $(1/32)^2$.

The SNR due to quantisation, $(S/N)_\text{A/D converter}$, can be calculated accordingly [3]:

$$(S/N)_\text{A/D converter} = 10\log_{10}(39^2 S) = 47 \text{ dB} \tag{9}$$

Comparing this quantisation SNR with the SNR due to receiver noise shows that the A/D converter has a negligible influence on the accuracy of the coarse ranging process.

Digital processing   After the analogue-to-digital conversion of the coarse ranging signal, further circuitry takes care of the processing. This digital hardware consists of a RAM and a microprocessor. A software program is loaded into the microprocessor to perform the processing on the ranging samples stored in the RAM. The objective of the processing is to obtain a coarse ranging phase delay value with the required accuracy. The processing input signal consists of the digital samples of the rather noisy coarse ranging sine wave. The phase measurement is based on discrete fourier transformation (DFT). Applying the DFT for phase determination is rather simple and straightforward. The DFT definition formula can be split into real cosine terms and imaginary sine terms.

Performing a DFT on a sine wave with a certain random phase ϕ will lead to a series of terms containing, after summation, real terms with $\cos(\phi)$ and imaginary terms with $\sin(\phi)$. From these terms the phase of the ranging signal is determined. Performing a DFT for one coarse ranging signal cycle will result in an inaccurate transform because of the noise on the ranging signal samples. This leads to an inaccurate coarse ranging phase delay value. Two different methods are possible for improving the accuracy. One method is to perform digital filtering before the DFT phase measurement action. Computer simulations have been performed, which show that the application of such a filtering method leads to inaccuracies due to harmonic distortion. The laser in the ONU transmitter introduces such a distortion and therefore this method cannot be applied. A second method is to perform the DFT phase measurement action over a large number of coarse ranging periods. In this way, phase measurement and filtering are performed by one algorithm. Simulations have shown that this method is insensitive to harmonic distortion. These simulation results have been compared with actual measurements on distorted coarse ranging signals, showing full agreement between simulation and measurements. Therefore this second processing algorithm is implemented in the BAF system.

It can be shown that the obtained equivalent noise bandwidth B_N, performing a phase measurement by a DFT over m periods of the ranging frequency f_r, can be approximated by:

$$B_\text{N} = \frac{f_\text{r}}{m} \tag{10}$$

The output noise signal after digital processing $\langle i^2_{\text{noiseout}} \rangle$ is the spectral noise density $S(f)$ according to (8) multiplied with the filter bandwidth B_N of (10). Taking the sinc value 1 for $S(f)$ gives:

$$\langle i^2_{\text{noiseout}} \rangle = S(f) \times B_N = \frac{\langle i^2_n \rangle f_r}{f_s m} \tag{11}$$

The CR signal has a value of $\langle i^2_{\text{ran}} \rangle$. The output signal-to-noise ratio SNR_{output} is:

$$SNR_{\text{output}} = \frac{\langle i^2_{\text{ran}} \rangle m f_s}{\langle i^2_n \rangle f_r} \tag{12}$$

The SNR_{input} is $\langle i^2_{\text{ran}} \rangle / \langle i^2_n \rangle$ and has a measured value of about 7 dB. The filtering improvement is the relation between output and input SNR:

$$\frac{SNR_{\text{output}}}{SNR_{\text{input}}} = \frac{m f_s}{f_r} \tag{13}$$

With a sampling frequency f_s of 1.39 MHz, and a ranging frequency f_r of 9.25714 kHz, the SNR improvement with $m = 1$ is f_s / f_r: 150 or 21.7 dB. Expressed in dB, the relation between the SNR_{output} and SNR_{input} of the DFT phase measurement over m periods is given by:

$$SNR_{\text{output}} - SNR_{\text{input}} = 21.7 + 10 \log_{10}(m) \text{ dB} \tag{14}$$

This formula gives the SNR improvement due to digital processing in the BAF system.

SNR_{input} has a value of 7 dB (Chapter 3.5.5.2). Owing to the CR accuracy requirement, the value of SNR_{output} has to be 44 dB (Section 3.5.4). This results in a DFT requiring a minimum value of $m = 32$ periods. Since the RAM can be filled with a maximum of 54 ranging periods, $m = 54$ is used. This means that the required accuracy can easily be obtained.

Measurements have been performed on prototype equipment to check the theoretical result given above. At the lowest CR level of −40 dBm at the receiver input, a CR standard deviation of 0.0035 radians has been measured. The requirement, calculated in Chapter 3.5.4, has been a value of 0.0066 radians. The obtained result is somewhat better than the requirement, which is caused partly by using more periods than required. The result clearly confirms that the calculations are reliable and that an implementation of this method of coarse ranging is possible. The statistical probability density function of the coarse ranging result has also been measured. This clearly shows the standard normal or Gaussian distribution function. So the assumption on which the CR accuracy requirement was based is correct. Measurements have been performed on crosstalk effects and show a negligible effect compared with the noise induced standard deviation of the phase measurement.

When the actual delay time measurement is completed, the OLT transmits an OAM message to the target ONU with the delay line setting for that ONU. With this setting, the total round trip delay equals the predefined delay, which is based on a 10 km network length, with an accuracy of $+/-0.4$ cell slot time and a 99% probability of success.

The static fine ranging and dynamic fine ranging procedures, as described in Chapter 3.5.3, will be carried out next. This results in an operational BAF system, with cells placed in their allocated slots within plus or minus one bit time.

REFERENCES

[1]   Gardner F M, *Phaselock Techniques*, 31–2, John Wiley & Sons, 1979
[2]   Carlson A B, *Communication Systems*, p 163, MacGraw Hill, 1986
[3]   Carlson A B, *Communication Systems*, p 435, MacGraw Hill, 1986
[4]   Dixon RC, Spread Spectrum Systems, John Wiley & Sons, 1976.

3.6 MAC PROTOCOLS FOR PASSIVE OPTICAL NETWORKS

3.6.1 Introduction

PONs are by their very nature shared-medium networks and there needs, therefore, to be some form of access protocol (the medium access or MAC protocol) that allows a station wishing to transmit to do so without causing any interference to other stations. This is exactly the same as the rationale behind the well known MAC protocols for LAN systems such as Ethernet or Token Ring. However, there is a considerable difference in topology between PONs and conventional LANs:

- there is a separate channel (fibre or wavelength) in each direction;
- the need for access control is different in the two directions.

In the downstream direction there is no access control mechanism since all the transmissions are from a single station, the OLT; it is in the upstream direction that control is required.

For the purposes of this discussion it can be assumed that ranging in the physical layer enables the time taken for a piece of information to be transmitted from any one node to the OLT to be known accurately. Indeed, the greatest simplicity is achieved if the ranging makes all nodes appear to be at exactly the same distance from the OLT so that the time taken is identical for each node. With PONs intended for narrowband networks, for CBR traffic only or for cases where the bandwidth demand of any station is small compared with the total bandwidth of the PON transmission channel, it is possible to use a simple TDM protocol, allocating regular time slots to each node in the upstream direction. However, with a PON intended to exploit the principles of ATM by providing a flexible bandwidth allocation to each node, allowing each to request as much bandwidth as is available subject only to normal dimensioning constraints, a more sophisticated protocol is required. This chapter describes the principles behind such protocols and then considers some of them in greater detail.

3.6.2 General protocol considerations

In the downstream direction ATM cells are sent by the OLT to every T interface with the address of the intended end-point being prepended to the cell in a tag. Since every interface will see all the cells transmitted, there is obviously a privacy problem, but this can be solved by encryption techniques. Such techniques have been implemented in the BAF system and are discussed in Chapter 3.8. In the upstream direction matters are somewhat more complicated because any number of T interfaces may wish to send cells at the same time. Hence there needs

Access to B-ISDN via Passive Optical Networks. Edited by U. Killat
© 1996 John Wiley & Sons Ltd

to be some form of medium access (MAC) protocol to ensure that only the information from a single interface is on the BAF system at any one time: collisions are not allowed.

Given the ability of ATM to cope with variable bit rate (VBR) services, it is important that the MAC protocol can also cope with such services and allow any customer to dynamically request bandwidth. Its performance, especially with respect to transfer delay and cell delay variation must be very good: the better the performance the higher the load that can be allowed and thus the higher the utilisation of the system. The MAC protocol must also be autonomous, i.e. the MAC protocol should rely only on ATM layer information that can be easily made available at connection set-up and it certainly should not be dependent on a particular type of local exchange. In this way the PON can be dropped in instead of a direct fibre connection.

The choice of the MAC protocol has a great influence on the performance of the system and a wide range of factors had to be considered. These are described and then details of the actual protocols devised for the BAF system are given. Considering protocols that use permit bits to allow upstream access there is a number of variations possible in the way that the permit bits can be transported:

• prepended to each downstream cell in a tag: this allows one upstream cell per permit;

• a tag in front of a block of cells: to allow a block of upstream cells to be together;

• in a fixed position in a frame.

The target of the permit can be the VCI, the VPI or a particular customer (T interface) or traffic class (VBR or CBR traffic, for instance). The choice of the target is a critical issue because it has particular influence on the system complexity and on the shaping imposed on traffic passing through the system. For example, a MAC protocol that addresses individual VPIs is likely to be more complex than one addressing T interfaces and would imply traffic spacing for that VPI.

The unit of transfer upstream can be either single-cell or multi-cell blocks. As in other issues there is a trade-off here between efficiency and complexity. Sending upstream traffic as a block of cells, rather than as single cells, increases the efficiency because there is less physical layer overhead but it is likely to lead to bunching of cells as they appear at the OLT. Any bunching is likely to require the use of a spacer to bring the cell delay variation (CDV) within acceptable limits.

The permit allocation algorithm can attempt to distribute permits to transmit upstream information based on the overall connection parameters (established at connection set-up) or try to respond dynamically at a burst level or at a cell level. Request information: with the dynamic allocation of permits based on cell level information, information about the state of the traffic at each queue must be sent upstream; excessive delays in sending this information would lead to a greater likelihood of queue overflow and hence cell loss.

A variety of protocols were considered within the BAF project and these can be divided into first and second generation, and frame and cell based techniques. The first generation proposals were all considered in parallel early in the project and one of these was chosen (Protocol Choice: Chapter 3.6.9) for use in the demonstrator. However, detailed technical studies were carried out on *all* of the protocols and the work and experience led to a second generation. These protocols are all considered separately: the global-FIFO, McFRED frame based protocol and the dynamic allocation algorithm (DAA) are all first generation; the belt permit programming (BPP) protocol is second generation.

In what follows, that part of an ONU which handles the MAC protocol for a single T interface is referred to as a slave MAC protocol controller (SMPC) and its peer in the OLT as a master MAC protocol controller (MMPC).

3.6.3   Global-FIFO protocol

The protocol actually implemented in the demonstrator system is the global-FIFO and the reasons for the choice of this protocol are discussed in Chapter 3.6.9. This is a cell based approach where each SMPC sends request information (information giving the current queue length of that SMPC) upstream in protocol control information at the front of each upstream ATM cell from that SMPC. However, there is a fundamental problem in using such a technique in that an SMPC that is not transmitting upstream does not have an opportunity to send request informationand so is not able to start sending cells at the start of a burst. One way round that problem would be to have a separate low bandwidth control channel open for each SMPC, which could carry request information. However, such an approach would waste bandwidth overall and the channel for each SMPC would be of relatively low bandwidth so that the time to get an upstream slot to send information would be too long, hence increasing the probability of buffer overflow at that SMPC. The concept chosen was to use all the spare bandwidth to carry upstream information: instead of sending idle cells, a different type of cell slot was invented, the request block. This is illustrated in Figure 3.6.1.

Including the concept of request blocks guarantees a fast reaction on changing traffic situations. By issuing permits for request blocks during

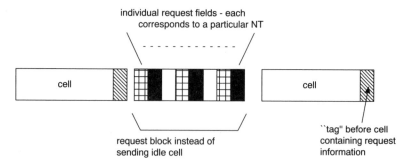

Figure 3.6.1
Cells and request blocks

idle periods of the global-FIFO queue, no bandwidth is actually wasted as otherwise no information would be carried. The request blocks used in the MAC protocol described here actually contain nine request fields each of which carries the request information for a different SMPC. If the BAF system is fully populated with 81 SMPCs then they can all be served within nine request blocks. Carrying multiple request fields in a request block complicates the physical layer but provides a very much better dynamic reaction performance; it also leads to lower buffer sizes in the SMPCs.

The permit distribution algorithm uses a concept called the global-FIFO where all request information is queued in one FIFO queue in the MMPC and permits are issued on a first-come, first-served basis across all the requests, for all the SMPCs. However, in order to reduce cell clumping, and the consequential increase in CDV, a spacing mechanism is implemented in the global-FIFO. This operates on a very simple principle: the MMPC keeps track of the total bandwidth allocated to each SMPC and will not issue permits to that SMPC at a rate greater than the peak rate corresponding to that total bandwidth. Because this spacing mechanism operates on the sum of the rates, rather than on individual VPs or VCs, it is named a bundle spacer. However, detailed work on the effectiveness of this bundle spacer has led to a modification called the bounded period rule. This is illustrated in Figure 3.6.2. The principle is that for lower bit rate sources, it is not necessary to space the permits by a period corresponding to the total bandwidth: permits can be allowed to be sent more frequently.

According to the bounded period rule the permit spacing period (of an SMPC), T_{sp}, can be related to the inverse of the peak bit rate (of an SMPC), T_{tot}, by the expression $T_{sp} = \min(\varepsilon T_{tot}, T_{max})$; with $0 < \varepsilon < 1$ and $T_{max} > 0$. T_{max} and ε are parameters defining the exact value of the bounded period rule and the values of these parameters are determined experimentally.

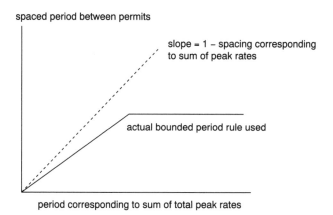

Figure 3.6.2
Bounded period rule

3.6.4   Frame based protocol

In the frame based access protocol (referred to as McFRED), the information flow is organised both in the upstream and downstream directions with frames of 125 μs; see [1] and [2]. The use of a framed approach allows a sampling of the SMPC status independently of the amount of traffic actually generated by the SMPC itself or by other SMPCs, and limits the maximum value of the reaction time. Each SMPC sends its request (MAC field: 8 bits) at the beginning of each upstream frame (Figure 3.6.3) and receives back the permits at the begin of the downstream frame (Figure 3.6.4). The request number corresponds to the number of cells arriving during the previous frame,

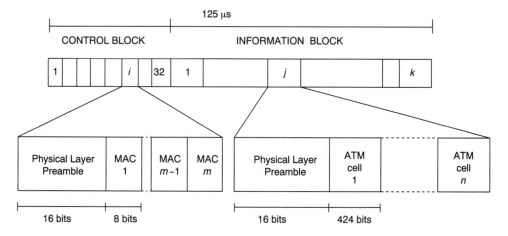

Figure 3.6.3
Structure of the upstream frame

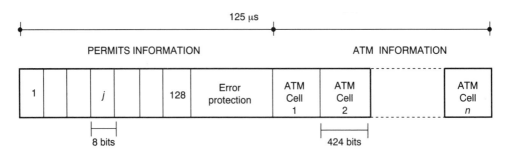

Figure 3.6.4
Structure of the downstream frame

while the permits number corresponds to the number of cells each SMPC is allowed to transmit. Both requests and permits are protected by a error correction field. Assuming the round-trip propagation delay of 100 μs, and 25 μs for the permit distribution algorithm (PDA) to receive, process the requests, and generate back permits, cells whose request has been processed in a frame could be transmitted in the following frame. Each SMPC, having the knowledge of the permits received by the previous SMPCs, is able with simple operations to evaluate the beginning of its transmission time. To minimise the number of preambles and then to maximise the link rate efficiency, cells transmitted from the SMPCs in the same ONU should all be concatenated behind the same preamble. On the other hand, to avoid traffic fluctuation of the SMPCs in the first positions to modulate the transmission of the cells of the SMPCs in the last positions, cells transmitted contiguously by the same ONU are limited by a value N_{max}.

The PDA evaluates permits both at connection level and at request level. The number of permits assigned at connection level, s_i (permits/frame) for the ith SMPC, is determined on the basis of the sum of the peak rates of the sources of the SMPC itself which have no benefits from statistical multiplexing (e.g. CBR sources and bursty sources with very high peak rate) and with a granularity derived from frame length. Permits at request level for the ith SMPC are distributed on the basis of the actual requests, t_i (permits/frame), and of the guaranteed permits in overload conditions, u_i (permits/frame). The u_i value is related to the amount of total bandwidth that the connection admission control (CAC) allocates for the ith SMPC.

The permits are determined in the following way. When the total requests (including preallocation and pending requests) are less than the available permits per frame, E_{tot}, all the requests are accomplished. In the opposite condition, when the total requests exceed E_{tot}, the permits are allocated according to these steps:

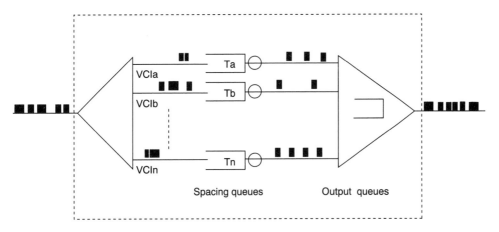

Figure 3.6.5
Internal architecture of the spacing unit

step 1:   permits for the ith SMPC are allocated according to:
$\min[u_i, \max(s_i, t_i)]$.

step 2:   at the end of step 1, some requests may remain unsatisfied.
If additional permits are available, the algorithm initiates a
cyclical distribution starting from the SMPC following the one
that has received permits in step 2 of the previous frame.

The framed approach may produce clusters of cells at full rate
(622.08 Mbit/s), heavily modifying the traffic profile generated by the
source particularly the peak bit rate. This could produce cell discarding
in the successive UPC/NPC even if the user is respecting the declared
peak bit rate. Moreover, the characteristics of the traffic submitted
to the ATM network elements become more bursty with a possible
degradation in the cell loss behaviour. To avoid these effects, the MMPC
is provided with a Spacer Unit. This element rebuilds the original peak
bit rate of cells to close together by adding extra delays, and operates
very similarly to a spacer policing unit defined in [3]. This unit is
based on a set of queues, one for each VC/VP served at the VC/VP
peak bit rate (Figure 3.6.5). In addition, an output buffer is added to
resolve the contention for the same output slot by several queues. This
logical subdivision could be realised using a common shared memory
as explained in [4].

3.6.5   The dynamic allocation algorithm protocol

The dynamic allocation algorithm (DAA) is an autonomous
request/permit based MAC protocol. The SMPCs transmit their

requests to the MMPC which distributes the permits to transmit ATM cells according to the needs of the SMPCs. The DAA issues permits to SMPCs based on the cell arrival rates at the SMPCs (although the absolute number is transmitted upstream for robustness reasons: see Chapter 3.6.8). In fact the DAA tries to match the permit arrival rate to the cell arrival rate so that the ATM cells leave the SMPCs at the same rate as they arrived at the SMPCs.

At the MMPC the cell arrival rate is derived from the received queue length. By keeping track of new cell arrivals and the time between these cell arrivals, the cell arrival rate at the SMPC can be estimated by the MMPC. The DAA algorithm uses a running window with check-points to achieve this (see Figure 3.6.6). Whenever the window crosses a check-point the MMPC checks to see whether there has been any cell arrival at an SMPC. If so, the bandwidth request for that SMPC is calculated by dividing the number of arrivals by the window size (i.e. value of the check-point) and the running window is then restarted. If however, there were no cell arrival the window would run until the next check-point.

The first check-point is called the W_{min} and so after the start of a window the protocol waits W_{min} slots before it checks for the cell arrival. If there is no arrival then it checks again at the next check-point. This continues until the last check-point, W_{max}, when if there is still no cell arrival the required bandwidth is set to zero and the running window is restarted.

The estimated cell arrival rate for each SMPC is stored in a bandwidth request table. The bandwidth request value can range from 0 (no bandwidth is requested) to 9360 (the full link capacity is requested). The figure 9360 is chosen to fulfil the 64 Kbit/s granularity requirement at a 599.04 Mbit/s capacity link. The permit distribution algorithm in the DAA protocol distributes permits according to the proportion of the values stored in the bandwidth request table. Since these values are a measure for the requested bandwidth to empty the SMPC buffers, the resulting permit rate matches the requested bandwidth. Ideally in

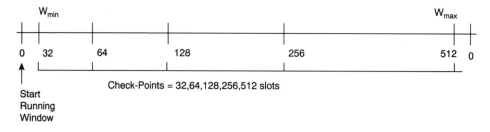

Figure 3.6.6
Example of the DAA algorithm running window

the case of no round-trip delay, the permit rate follows the cell rate arriving at the SMPC accurately.

The DAA permit distribution algorithm can be compared to a horse race where SMPCs contend for a permit. Each SMPC races with the speed of its bandwidth request value and each time slot is won by the SMPC that lies ahead. An SMPC with a high bandwidth request races faster than an SMPC with a small bandwidth request and is thus served more often. Besides the speed of the SMPC the position of the SMPC in the race is used. The speed is determined by the value stored in the bandwidth request table, while the position is maintained in the urgency table. As the name implies, the urgency table is a measure of the need of an SMPC to get a permit.

In general the sum of the required bandwidths is less than the available link capacity. The average load in an ATM system may be in the order of 0.8. This means that there is generally some spare capacity. Originally the DAA protocol distributed this spare capacity proportionally to the bandwidth requests among the SMPCs. This way the SMPCs got more bandwidth than actually requested which could be used to transmit more request messages resulting in a faster reaction time. However, the concept of request blocks where multiple SMPCs can transmit their request in the same time slot appeared superior. Therefore this concept was included in the next version of the DAA. Request blocks can easily be added to the DAA by introducing a virtual SMPC that claims the spare capacity. In fact the bandwidth requested by the virtual SMPC is the complementary value of the sum of the requested bandwidths of the other SMPCs. The permit distribution algorithm now also generates permits for this virtual SMPC corresponding to the spare capacity. Each time a permit is generated for this virtual SMPC, a permit for a request block is issued by the MMPC. Now the SMPCs get exactly the bandwidth they requested, while the spare capacity is used for request blocks.

The pseudo-code for the DAA permit distribution algorithm is given below. The urgency of the virtual SMPC is set to zero. Then, the urgency values of all SMPCs are compared with each other. The SMPC with the highest urgency value gets the permit (the horse at the head of the race wins). In the mean time the urgency of the virtual SMPC is calculated such that the sum of all urgencies is zero. Moreover the urgency values of the SMPCs are incremented by their bandwidth request values (the horses race according to their bandwidth request value). Finally the urgency of the virtual SMPC is compared with the urgency of the winning SMPC. If the urgency of the virtual SMPC is not lower, a permit for a request block is issued, else a permit for the winning SMPC. If a permit is issued for a SMPC, then the urgency is decreased by the maximum bandwidth value (the horse is put back in the race).

```
UrgMax := -BWmax
Urgency(VSMPC) := 0
for all (SMPCs) do {
    if (Urgency(SMPCi) > UrgMax) do {
        UrgMax := Urgency(SMPCi)
        permit := SMPCi
    } endif
    Urgency(VSMPC) := Urgency(VSMPC) - Urgency(SMPCi)
    Urgency(SMPCi) := Urgency(SMPCi) + BandwidthRequest(SMPCi)
} endfor
if (Urgency(VSMPC) >= UrgMax) do
    permit := VSMPC
else
    Urgency(SMPCi) := Urgency((SMPCi) - BWmax
endif
```

Figure 3.6.7
Pseudo-code of the DAA permit distribution algorithm

Table 3.6.1
Example of the DAA permit distribution algorithm operation

SMPC	B_{req}	Urg_0	Urg_1	Urg_2	Urg_3	Urg_4	Urg_5	Urg_6	Urg_7	Urg_8	Urg_9	Urg_{10}	Urg_{11}
A	3	0	3	−4	−1	2	5	8	1	4	7	0	•
B	1	0	1	2	3	−6	−5	−4	−3	−2	−1	0	•
C	1	0	1	2	3	4	−5	−4	−3	−2	−1	0	•
D	2	0	2	4	−4	−2	0	2	4	−4	−2	0	•
vnt		0	−7	−4	−1	2	5	−2	1	4	−3	0	•
permits		RB	A	D	B	C	RB	A	D	RB	A	RB	•

An example of the operation of the DAA permit distribution algorithm is shown in Table 3.6.1. In this example four SMPCs are present called A, B, C and D. The fifth SMPC is the virtual SMPC that takes care of the generation of request blocks. In the example the total bandwidth is divided into 10 units (instead of the 9360 units in the BAF system). The bandwidth request values are chosen such that of the available 10 units in bandwidth SMPC A requests three units, SMPCs B and C both request 1 unit while SMPC D request 2 units. Since the total requested amount of bandwidth is seven units, the virtual SMPC requests the last three units. At $T = 0$, all SMPCs start with an Urg 0. Table 3.6.1 shows how the urgencies of the SMPCs change and how the permits are generated for the SMPC with the highest urgency value. As long as the values in the bandwidth request table do not change, the pattern repeats itself every ten slots. However, in a flexible environment like ATM the values in the bandwidth request table may change with every request update. The permit distribution algorithm adjusts the urgencies accordingly and new permit rates will be generated automatically.

3.6.6 The belt permit programming MAC protocol

The dominance of the round-trip delay in the reservation based access methods for the tree ATM PON greatly affects the performance of the upstream multiplexing/concentration function. The effect of the MAC is to increase the peak rate of the connections and make the traffic appear more bursty. The increased CDV in conjunction with the preventive congestion control of ATM results in reduced efficiency. A jittered flow consumes more link capacity than the original unjittered cell stream in the sense that the connection admission control (CAC) of ATM can accommodate much less total traffic if jittered. Consequently the CDV behaviour of an access arbitration protocol acquires prominence.

The MAC protocol presented below places the emphasis on lowering the CDV of the stream exiting the OLT to levels approaching those expected from an ideal multiplexer. To this end it exploits the observation that a global-FIFO scheduling is possible thanks to the electronic distance equalisation in the system. A downstream frame transmission provides a time marking which is of global value, i.e. all ONUs perceive it simultaneously when the electronic delay obtained through ranging is also taken into account. Thus the scheme presented below makes use of the arrival timing within a frame and presents this more detailed information in the request field instead of just the number of arrivals, thus enabling the MAC controller to exercise a true FIFO scheduling of the permits.

3.6.6.1 *Transmission format*

Transmission in both directions is organised in implicit frames of 144 slots. Each downstream slot (Figure 3.6.8) contains an ATM cell of 424 bits preceded by a control field (CF) of 16 bits containing the permit in the form of a 5-bit control address identifying the SMPC. In the upstream frame (Figure 3.6.9) each cell is preceded by the physical layer preamble which is indispensable for bit and byte alignment at the head-end. However every 17 slots a whole slot, called a request access unit (RAU), is dedicated to providing control information to the MAC controller in the OLT (MMPC). Each periodic RAU is formed from emissions from several terminations. Since the rate of the termination is one quarter that of the system, each termination can at most have only one arrival every four slots of the 622.08 Mbit/s system.

In the case of FTTC every three terminations (at most) multiplex their traffic at the ONU into an entity called a multiplexed unit (MU) and present one common request through one common SMPC. Multiplexing in this way does not handicap terminations which belong to groups as opposed to those enjoying an MU to themselves because

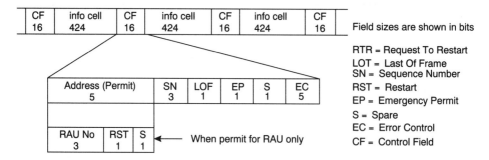

Figure 3.6.8
Downstream transmission format for BPP

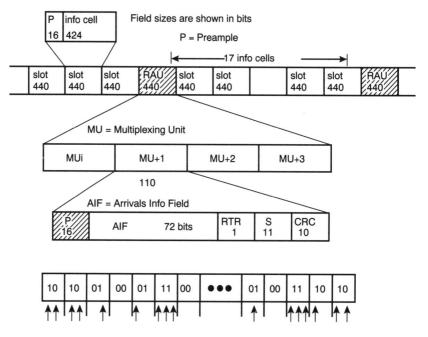

Figure 3.6.9
Upstream transmission format for BPP

the criterion for a permit is the arrival time. ONUs supporting more than three terminations will have more than one MU and up to four MUs. This PON system can support at maximum 96 terminations configured into 32 MUs and 32 ONUs. For every ONU with more than one MU, the total number of required ONUs is reduced correspondingly down to just eight if all are fully loaded with four MUs (i.e. 4 SMPCs, supporting 12 terminations). In the other extreme configuration of only FTTH, the MU and ONU become one component. The corresponding arrival information field (shown in Figure 3.6.9) then refers to a single termination.

With the help of the implicit frame of 144 cells each termination can mark the exact time of arrival (with a resolution of one 155 Mbit/s slot) of each cell relative to the frame. As the start of the frame is considered the start of the RAU which has been allocated by configuration management the value RAU0. The arrival information which is relative to the start of frame is conveyed to the MMPC. As one RAU is issued after 17 payload slots, all 32 MUs are covered in the eight RAUs of a frame. Each MU is given a 72-bit long field in the RAU for announcing arrivals in the following way. The 144 slot period is divided into 36 segments of four slots each. The number of arrivals in each segment is indicated by using 2 bits. (Maximum number of possible arrivals in each MU is three). Thus 72 bits cover all 36 segments and allow for even the extreme case of 144 arrivals in one MU. Each arrival implies a request for an upstream permit. In addition 1 bit is employed for enhanced robustness, 11 bits marked as spare are reserved for other priorities, and 10 bits are used for CRC.

3.6.6.2   The permit allocation algorithm

The permit allocation method will be described with the help of Figure 3.6.10. It is based on the reading and writing of a fast RAM which will be called FIFO permit multiplexing RAM (FPMR). This RAM is depicted in Figure 3.6.10 as a conveyor belt carrying the scheduled permits in a metaphor aimed at assisting the description of the workings of the scheme. The permits are the addresses of the MUs

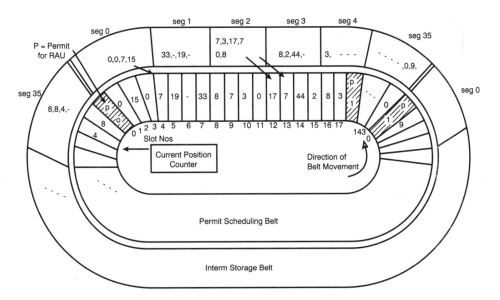

Figure 3.6.10
The FIFO permit multiplexing RAM (FPMR)

(or ONU-Hs in case of FTTH) and are emitted towards the terminations inside the pertinent field of the downstream cells.

As each RAU arrives at the OLT the arrivals during each segment are made known to the MMPC and they are stored in the interim storage belt. So, if for example the first MU with address 0 had two arrivals in the first segment (seg0), the address 0 is written there twice. If then it had an arrival in another segment (e.g. seg2 and seg35), its address is written there too, as in all other segments corresponding to cell arrivals to that MU. After the arrivals of MU0 for the whole cycle have been recorded, the arrival information of MU1 is then stored in the FPMR in a similar fashion. The process is repeated with each arriving RAU until all RAUs in a cycle have been thus stored in the Interim Belt and a new cycle starts with RAU0 again. Thus, a reflection of the arrival status at the terminations is continuously created (and overwritten by later information) in the interim storage belt.

To avoid favouring the first MUs in a RAU, arrivals are not immediately transferred into the permit scheduling belt, but only after all the requests from all the RAUs of a cycle have been filled into the interim storage belt. All permits of each segment are registered before those of the segments which follow in accordance with the FIFO service principle. Within the same segment all permits are considered equivalent as there is no resolution of arrivals within a 155 Mbit/s slot. Blanks (shown as dashes in Figure 3.6.10) resulting in empty permits, are transferred only if no other permits from previous segments are outstanding.

Simultaneously and while the next frame is being prepared using upstream arriving RAUs, permits already scheduled from the previous frame are emitted with every downstream slot. The current position counter marks the one to be sent next. In the long run the rhythm of reading the FPMR, which is dictated by the downstream frame, and the writing of the FPMR, which is tied to the periodic arrival of RAUs, are synchronised with a difference of one propagation delay (i.e. when the last permit of a downstream frame is sent, the last RAU of the corresponding upstream frame has been received and processed so that the next frame of the FPMR is ready).

If the collisions of cells arriving in the same slot are ignored for a moment, the delay of all cells is fixed at one frame plus one round-trip delay. (The one frame delay comes from summing the time from arrival till the respective RAU is sent plus the time the cell has to wait from the scheduled position till the permit is sent. The round-trip delay experienced by all cells is due to the propagation time of the RAU plus the propagation of the permit to the termination). Of course when more than one cell arrives at the same slot, all but one will have a higher delay and also push further the cells arriving on the following slots.

But then this happens with any slotted multiplexer and is the cause of the unavoidable residual jitter.

The end result is that with the help of the information of the RAUs, the MMPC creates in the interim storage of the FPMR a reflection of the queue status at the SMPCs including inter-arrival timing information. Thus the stream exiting the OLT is not very different from the one which would come out of a concentrated slotted FIFO multiplexer albeit with an additional fixed delay equal to one frame plus one round-trip propagation delay (just over 200 microseconds). Regarding CDV, the difference is only that due to the unavoidable waste for the RAUs, which amounts to $\frac{1}{18}$ of the link capacity and the lack of detailed resolution. There is also of course the additional overhead due to the 16-bit tag in each slot of 440 bits. However this overhead does not affect the CDV but only the efficiency by increasing the size of the slot, not the useful bits.

Summing up, the belt protocol exploits arrival timing information to produce a traffic stream which conforms to the dictates of the ATM policing and connection acceptance philosophy, while maintaining functional decoupling from the rest of the network. No call related functions are invoked.

3.6.7 Robustness

3.6.7.1 Robustness in MAC protocols

The MAC protocol of an access network must be able to work under non-ideal conditions. Owing to synchronisation failure, noise or other reasons, there can always be bit errors (detectable or undetectable) or cell loss in the information received. Adding overhead and using an error correcting code can reduce the error probability but there is always a residual probability that an error is not detected or is wrongly corrected.

MAC protocols relying completely on a reservation principle, like the request–permit mechanisms of the protocols presented in this book, must be able to cope with errors like loss or distortion of control information. If no precautions are taken, such errors can severely affect the delay performance of a system. For example, when the information about a cell arrival gets lost, the next permit is used to transmit the affected cell and all following cells will be delayed accordingly. The last cell of a connection could even be stuck in the SMPC queue for a very long time. Such situations would have to be detected by performance monitoring using OAM cells, which takes a very long time until detection, or creates much overhead. After detection, the corrective actions to be taken for recovery, like a complete MAC reset, would additionally deteriorate the MAC performance. Therefore, a robustness mechanism is built into the MAC protocols which prevents errors in the MAC control information from having long-term and accumulative effects.

There are two basic cases of distorted control information:

a) The request shows too many arrivals. This results in worsened CDV behaviour of the MAC causing cell loss at the network parameter control (NPC) later in the network and in the assignment of permits to an empty SMPC queue causing this SMPC to transmit idle cells.

b) The request shows too few arrivals. In this case there are two alternatives for recovering from the resulting lack of permits at the affected SMPC, corresponding to two protocol optimisation goals which reflect different kinds of service requirement.
   - the cell loss is minimised. Additional permits are generated when the situation has been detected;
   - the CDV is minimised. Cells which are not covered by the request received are discarded at the SMPC queue.

The globals FIFO and BPP MAC proposals differ in this last respect, as can be seen below.

3.6.7.2   Robustness mechanism of the global-FIFO protocol

The global FIFO robustness mechanism follows two general rules:

a) Redundant information is used for coding the requests. Instead of transmitting the number of new arrivals in a request field, the current SMPC queue length (i.e. T interface queue length) is transmitted and the number of arrivals is derived from this information.

b) It prevents MAC-induced cell loss by trying to be on the safe side when issuing permits. This is achieved by choosing an algorithm which generates too many permits rather than too few if there is inconsistency in successive requests from the same SMPC.

Figure 3.6.11 shows the principal steps in serving an SMPC when a request has been received.

The mirror counter scheme tries to make the number of permits in the system for one SMPC, say $SMPC_1$, equal to the number of cells in the system from the same SMPC. The number of permits is reflected by the value stored in the mirror counter. Figure 3.6.11 shows that whenever request information (that is the queue length Q_i) is received from $SMPC_i$, Q_i is compared with the last known M_i value stored in the mirror counter for this SMPC and for the difference, permits are generated. M_i is then increased to the latest value of the SMPC queue length (if applicable, according to $M_i = \max(Q_i, M_i)$), but never decreased at

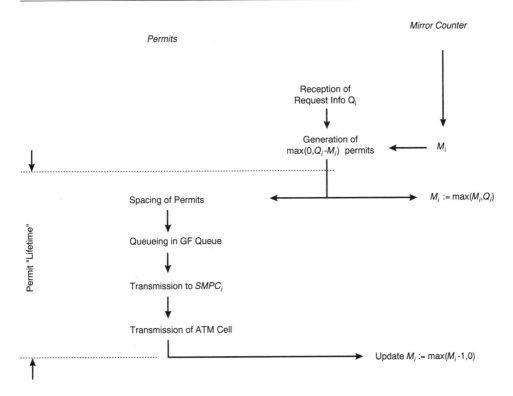

Figure 3.6.11
Principal steps in serving an SMPC when a request has been received

this step. Now instead of queueing a new permit directly, it is stored by incrementing the request counter RC_i and spaced (see the description of bundle spacing) before it enters the global-FIFO permit queue. When the permit is received by $SMPC_i$, an ATM cell is transmitted upstream to the MMPC (i.e. OLT). When the time arrives that the ATM cell is expected from $SMPC_i$, M_i is reduced by one, independent of whether an ATM cell or an idle cell arrived, or an error occurred. M_i is never reduced below zero. Figure 3.6.12 gives an example of the operation of the mirror counter scheme.

3.6.7.3   Robustness of the mirror counter scheme

The reaction of the mirror counter scheme to transmission errors can be characterised by looking at the following cases.

(a)   Request information is lost or its value Q_i is too small.
- no permits are "taken back" from the request counter;
- too few permits are generated initially if there have been new arrivals. When the next correct request is received, the missing permits are generated;

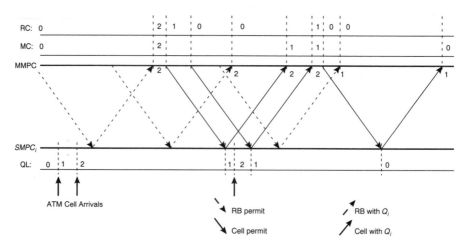

Figure 3.6.12
Example of the operation of the mirror counter

- a number of cells is delayed by the time between the distorted and the next correct request.

(b) Q_i is too large.
- too many permits are generated and no more permits are generated until $M_i < Q_i$ for the next time;
- cells arriving before the excess permits have arrived at the SMPC experience a lower delay, which may result in an increase of CDV.

(c) A permit is lost downstream.
- the mirror counter is decremented when the corresponding ATM cell is expected at the MMPC;
- the following request still shows one cell more than expected, which results in the generation of a new permit for the affected SMPC;
- the effect of this error is limited to 100 microseconds+reaction time.

(d) Loss of an upstream ATM cell
- this is the normal case of cell loss as in any other transmission system;
- the request prepended to the ATM cell is lost, see case (a).

(e) Change of permit destination
- this is a mix of cases (c) for the SMPC to which the permit was assigned and (b) for the SMPC which gets the excess permit.

The above summary shows that the mirror counter scheme is a suitable means of removing the long-term effects of transmission errors in control information from an MAC protocol.

3.6.7.4   Robustness of the belt permit programming protocol

The excellent CDV behaviour of the belt protocol is conditional upon a proper alignment between the permits and the corresponding cells, given the importance of permit timing. The provisions guaranteeing robust behaviour in the presence of the inevitable transmission errors in the control information fields will be described below. The actions adopted include error detection fields covering the control information, a demarcation of the cells of each period, and a restart mechanism for recovery from any error condition.

As shown in Figure 3.6.9, the permit is protected by a 5-bit error control (EC) field. By employing Hamming coding for hardware simplicity in the permit recognition circuits, several-bit error detection and one-bit error correction is possible. In the event of unrecoverable loss of a permit, a misalignment could set in for a long sequence of permits. This is avoided by a sequence numbering of permits (SN). Thus with the next permit the relevant MU will drop a number of waiting cells equal to the lost permits. This SN refers to each MU and uses 3 bits allowing detection of up to seven lost permits (cf. with AAL 1). Any greater disruption is considered a major event leading to a restart of the protocol initiated by the demarcation mechanism as described below.

Regarding the upstream control flow, the requests in the form of MUs are protected by a 10-bit CRC. The algorithm specified in recommendation I.610 is suitable. The extra circuitry is well justified and only centrally used once in each OLT. When loss of a whole request unit (RAU) is detected, the OLT continues with the following set of actions. Not knowing the number of cells waiting at the MUs of the lost RAU, it sends emergency permits (bit EP set) for these MUs whenever there is an unoccupied position in the permit scheduling belt. The four MUs are served in sequence and if the loading conditions allow, it is often possible that all the cells of the lost RAU may depart although not at the right positions according to arrival timing. Thus in this case the belt will simply offer a reduced CDV performance. If however loading is rather heavy there will not be enough empty permits to allow emission of all outstanding cells, which will then purged the moment the first proper scheduled permit arrives (intended for the first cell which belongs to the next correctly received RAU).

Finally there is still the case, though rare, of residual CRC errors, (i.e. cases where the protocol believes an errored AIF field to be correct). Whatever the source of misalignment, the protocol will realign itself without resorting to management plane action. This self recovery at the MAC level is based on the continuous demarcation of the completion of service of the cells in each frame. The tool for this is the bit LOF (last of frame) in the permit field (see Figure 3.6.8). This bit marks the last permit in each service cycle (frame) for each RAU. A corresponding tracking mechanism in the ONU marks the cells which belong to each reservation cycle. Therefore when the last permit is expected by the ONU it must be matched by a permit with the LOF bit high; i.e. the last permit in the frame for that RAU. In the event of mismatch the ONU sends a request to restart.

The restart mechanism (also employed for a first-time initialisation) is foreseen as a last resort of any failure of the protocol. It is based on the bit marked RST bit (ReStart) which is the fourth bit in a permit for RAUs (the first 3 bits in this field identify redundantly the RAU number). When RST is asserted, no permits should be considered as outstanding for this MU and therefore whatever arrivals will be registered in the next MU are the only ones to receive permits. The restart can also be initiated by the ONU using the bit RTR (request to restart).

With the demarcation and restart mechanisms the protocol keeps a tight watch over any misalignment and restarts with the loss of the information contained in the disrupted frame only. An initial restart (global restart), is effected by just marking an RST command in all RAUs.

3.6.8 Protocol choice for the demonstrator

In 1992 a decision had to be taken on a protocol to be implemented in the BAF demonstrator. At this stage, all of the first generation protocols had been devised but only early performance studies had been carried out; the decision had to be based mainly on engineering judgement. The results of the simulations that evaluated these decisions are given in the next chapter.

The factors leading to the decision on the MAC protocol for the demonstrator were:

- in order to achieve maximum performance it was decided to use a cell-level algorithm based on the lengths of the queue at each T interface;

- there should be no separate spacer, and as the CDV introduced by a frame based protocol without spacing was thought to be excessive, the protocol would be cell based;

- there should be *one* upstream cell per permit as this has the greatest potential for minimising the disturbance to the traffic profile without the inclusion of a separate spacer;

- although the BAF system was based on VPs the target of permits was chosen to be each T interface, which implies that bandwidth was allocated to the *bundle* of VPIs and VCIs originating from each interface.

These criteria could have been satisfied by the DAA or by the global-FIFO permit distribution algorithms but the initial performance studies suggested that the global-FIFO would lead to a lower CDV being introduced and hence that was chosen.

REFERENCES

[1] Delli Priscoli F, Listanti M, Roveri A and Vernucci A, *Access and switching techniques in an ATM user-oriented satellite system*, Proc of INFOCOM '89, Ottawa April 1989

[2] Angelopoulos J D, et al, *Performance of shared medium access protocols for ATM traffic concentration*, ETT, **5**, (2), Special issues on Teletraffic Research for Broadband ISDN in the RACE programme, 1994

[3] Boyer P, *A congestion control for the ATM*, Proc 7th ITC Seminar, New Jersey, October 1990

[4] Wallmeier E and Worster T, *The spacing policer, an algorithm for efficient peak bit rate control in ATM networks*, Proc ISS93, Yokohama, October 1992

[5] Angelopoulos J D and Venieris I S, *A distributed FIFO spacer/multiplexer for access to tree APONs*, SUPERCOMM/ICC '94 Conference New Orleans, Louisiana, May, 1994

[6] Angelopoulos J D, *Time division sharing of tree PONs by ATM users: A method to control cell jitter*, EUROPTO (SPIE/EOS) Conference on Broadband Strategies and Technologies for Local, Metropolitan and Optical Access Networks, Amsterdam, The Netherlands, March 1995

[7] Garcia J, Casals O and Blondia C, *A cell based MAC protocol with traffic shaping and a global FIFO strategy*, RACE Open Workshop on Broadband Access, June 1993

3.7 PERFORMANCE EVALUATION OF MAC PROTOCOLS

3.7.1 Introduction

When a new system is designed, different alternatives offering the same functionality have to be compared with respect to performance, reliability, modularity, robustness, hardware and software complexity, development costs etc. In particular when no implementations are available, modelling is often the only way of evaluating and comparing the alternatives quantitatively. Modelling is not only concerned with resource dimensioning problems, but also leads to an assessment of the merits of different operational mechanisms and strategies. This makes modelling a powerful technique, suitable for use in the early stage of system design.

In a later phase of system development, when a first simplified realisation of the system is available (e.g. demonstrator, pilot system, field trial), the modelling assumptions have to be validated and possibly corrected and refined, on the basis of measurements obtained from experiments performed with the pilot system.

In Chapter 3.6 four different MAC protocols are defined, all using different methods and strategies to control the access of a shared link to an ATM network. The goal of this chapter is to compare the performance of these protocols with each other in order to assess the advantages of each individual protocol under different load and traffic conditions. This performance evaluation is achieved by simulating each of the four MAC protocols. A more detailed description of the simulation approach used and the difficulties which occurred during the simulation studies are given in Chapter 3.7.3. To evaluate the performance of the described MAC protocols, traffic must be offered to the BAF system. The used source models and traffic mixes are described and explained in Chapter 3.7.2. The evaluation of the performance is presented in Chapter 3.7.4 and is attained by studying the complementary distribution function of two important performance characteristics for telecommunications systems in general and for a BAF system in particular: the transfer delay (Chapter 3.7.4.1) and the introduced cell delay variance (Chapter 3.7.4.2).

3.7.2 Source models and traffic scenarios

Traffic with variable bit rates will become important for wide area networks in the future. The age of multimedia workstations is just at its beginning, since services with varying bit rate present the future of telecommunications: transmission of compressed video, transmission of high-resolution graphics (e.g. X-ray pictures) and other real-time services like video conferences or HDTV which future networks have to

Access to B-ISDN via Passive Optical Networks. Edited by U. Killat
© 1996 John Wiley & Sons Ltd

support efficiently. Besides these future services, an ATM network (and therefore the BAF system) must be able to support familiar services like telephone conversations and fax messages.

In order to capture the major traffic characteristics of the different teleservices, the following three models have been selected to evaluate the performance of the different MAC protocols:

- **CBR traffic** This source model generates cells with almost[†] constant inter-arrival times. Both the service bit rate and the AAL overhead of 1 byte (CBR traffic is assumed to use AAL layer 1, requiring 1 byte per cell) are taken into account in calculating the cell inter-arrival time. This source model captures most conventional and near-future teleservices such as telephone, (fast) fax, (low quality) videophone and LAN–LAN on demand.

- **On-off traffic** This source model generates cells according to an underlying Markov chain alternating between two states: "on" and "off". When the Markov chain is in the on state, cells are generated according to a CBR pattern, whereas during the off state no cells are emitted. Both the number of cells generated during the on period and the sojourn time in the off state are chosen to be geometrically distributed. This source model is characterised by three parameters: the peak bit rate, the mean bit rate and the burst length (i.e. the mean number of generated cells during the on period) and was proposed in [8] as a model for data traffic (e.g. flexible file transfer).

- **Worst-case traffic** This source model generates b cells on full link rate and then stops sending cells until its mean bit rate is reached. This cell pattern is generated repeatedly. Where necessary, the required parameters (e.g. for the bundle spacer of the global-FIFO) are computed as if the source generates cells according to a CBR pattern. This source model is used to investigate the influence of unexpected and extremely bursty traffic on the delay characteristics of simultaneously active CBR sources.

The used source models and traffic mixes are partly based on a forecast of possible services which are likely to become important for users of a BAF system. For a performance study of MAC protocols, however, it makes no sense to investigate only one complex traffic mix. For complex systems such as the BAF system a quantitative analysis often hides problems which might have an important impact on the performance of the system. Performance studies must therefore be more general in order to be "service transparent" to a high degree and in order to capture the impact on the performance characteristics

---

[†] Granularity, caused by the possibility that a service rate is not divisible by the rate of the access link, has to be taken into account.

of all different features of the four MAC protocols. Taking this into consideration, the traffic sources given in Table 3.7.1 have been chosen as input traffic to achieve the goal of this chapter. The homogeneous and heterogeneous scenarios used for the performance evaluation of all MAC protocols are given in Tables 3.7.2 and 3.7.3 respectively. The used bit rates in these tables are defined at the AAL level, i.e. overhead introduced by both AAL and ATM layers is **not** included yet. For Scenario S8 to S10 the CBR source is considered to be the *tagged* connection, i.e. the connection from which the performance measures are obtained.

Multiplexing gain   When a traffic scenario is so bursty that temporary overload situations may occur (this happens for example with a concentration of on–off sources), the statistical multiplexing effect of an ATM traffic concentrator such as the BAF system provides a multiplexing gain. The multiplexing gain of a concentrator is defined as the ratio of the sum of the peak bit rates of all connections and the allocated bandwidth. Hence, with peak bit rate allocation the multiplexing gain equals one. The accomplished multiplexing gain of the used traffic scenarios is included in Tables 3.7.2 and 3.7.3.

Table 3.7.1
Traffic sources used

Source Name	Source Type	Peak bit rate	Mean bit rate	Mean Burst Length	Number of AAL bytes per cell
Type A1	CBR	64 kbit/s	64 kbit/s	–	1
Type A2	CBR	2 Mbit/s	2 Mbit/s	–	1
Type A3	CBR	10 Mbit/s	10 Mbit/s	–	1
Type A4	CBR	34 Mbit/s	34 Mbit/s	–	1
Type B1	ON-OFF	10 Mbit/s	1 Mbit/s	10 cells	4
Type B2	ON-OFF	50 Mbit/s	5 Mbit/s	100 cells	4
Type B3	ON-OFF	50 Mbit/s	5 Mbit/s	500 cells	0
Type C1	WCT	155 Mbit/s	10 Mbit/s	$b = 5$ cells	1
Type C2	WCT	155 Mbit/s	10 Mbit/s	$b = 20$ cells	1

Table 3.7.2
Homogeneous traffic scenarios used

Scenario Name	Source	Number of Sources per ONU	Number of Active ONUs	Load	Multiplexing Gain
S1	A1	82	81	0.80	1.00
S2	A2	3	70	0.79	1.00
S3	A4	1	12	0.77	1.00
S4	B1	12	30	0.68	7.50
S5	B1	3	81	0.46	5.06
S6	B2	1	45	0.42	5.00
S7	B3	1	40	0.37	4.00

Table 3.7.3
Heterogeneous traffic scenarios used

Scenario Name	Active ONU	Source Types	Load	Multiplexing Gain
S8	0 ... 18	$1 \times A3 + 3 \times B1$	0.49	2.20
	19	$1 \times A3$		
S9	0 ... 8	$4 \times A3$	0.76	1.00
	9	$3 \times A3 + 1 \times C1$		
S10	0 ... 8	$4 \times A3$	0.76	1.00
	9	$4 \times A3 + 1 \times C2$		

Load   The load of the BAF system (denoted by ρ) is defined as the ratio of the mean link capacity of the ATM layer which is used by assigned cells and the net link capacity. Since the internal overhead and link capacity in the BAF system differs per interface, different values of the load exist at T_b, U and V_b interfaces.

Because the load at the U interface depends on the used MAC protocol and the load at the T_b interface only supplies local information, the load at the V_b interface is chosen as reference and is included in Tables 3.7.2 and 3.7.3. The load at the V_b interface (ρ_v) can be computed as follows:

$$\rho_v = \sum_{\text{Active } T_b \text{ interfaces}} \rho_t \cdot \frac{\text{Net } T_b \text{ capacity}}{\text{Net } V_b \text{ capacity}}$$

where

$$\rho_t = \sum_{\text{connections}} \frac{\text{Mean bit rate} \cdot \dfrac{53}{48 - \text{AAL bytes}}}{\text{Net } T_b \text{ capacity}}$$

and
Net T_b capacity $= 149.76$ Mbit/s.
Net V_b capacity $= 599.04$ Mbit/s.

3.7.3   Simulation tools

Besides a model describing the characteristics of the traffic offered to the BAF system (as given in Chapter 3.7.2), there is need for a system model which describes the BAF system and its resources in detail. Once the system and its load have been modelled, analysis methods have to be applied in order to derive the required results. Two analysis approaches can be considered: simulation and analytical methods. When using a simulation method, the behaviour of the system is described by a program and the traffic is generated using random number generators. During the execution of the simulation program, statistical data (e.g. buffer lengths, transfer delays) concerning the operation of the BAF system are gathered. These data are used afterwards to compute the required performance measures. Simulation is considered to be a powerful performance analysis technique since it allows an analysis of detailed models. Stable statistical results with respect to rare events (e.g. cell loss) may require enhanced mathematical techniques and may even be impossible to obtain (see [7] and [9]). Analytical models, on the other hand, treat less detailed models but allow one to obtain more stable results with a higher resolution. Moreover, in some cases they may lead to closed-form solutions, which may allow straightforward interpretation. Analytical methods for the performance evaluation of the BAF system are mainly based on queueing analysis techniques. Analytical evaluation, however, often suffers from huge state space requirements when dealing with a superposition of many CBR or (a few, but bursty) on–off sources. Since it is impossible to give a detailed analytical queueing description of the BAF system, simulation methods are used to determine the required performance measures.

Four different simulators of the BAF system have been built, each of them operating with a different MAC protocol. During the simulation studies two problems occurred:

(1) **Simulation of many low bit rate or very bursty connections** When the BAF system is loaded with many low bit rate connections, the time between two successive tagged cells is large while simulation resources (such as event lists) are used extensively. This causes long run times until satisfactory 95% confidence intervals are obtained, allowing a 5% margin. The same problem occurs when the input consists of more bursty connections with a relatively low mean bit rate with respect to the peak bit rate.

(2) **Periodicity of CBR sources** When a BAF system is only loaded with traffic sources which generate an almost constant pattern of cells for a long time (such as CBR sources do), the gathered simulation results strongly depend on the (random) time a source waits before it starts to generate cells. A completely correct solution to this problem would be to perform runs for all possible combinations of phase shifts between the sources. This number of runs grows exponentially with the number of offered sources, resulting in unacceptably long simulation times. In order to simulate the BAF system which is symmetrically loaded with many highly periodic cell streams, *quasi-periodic* cell streams are used instead. A quasi-periodic cell stream generates a periodic cell pattern for a large number of cells and then waits for a random time X

before it starts to generate its cell pattern again (repeatedly). By choosing the expectation value of the random variable X ($E[X]$) such that it equals the inter-arrival time between two cells of the original CBR source, an unbiased estimator for the bit rate pattern of the original CBR source is obtained. The variance of the random variable X should not be too high otherwise the arriving cell pattern might suffer too much jitter. On the other hand, this variance should not be too small either, otherwise the cell stream might still be too close to periodic. A good choice for the distribution of X seems to be a uniform distribution over the interval $[0.85 \cdot E[X], 1.15 \cdot E[X]]$. However, one remark must be made on the use of a quasi-periodic cell stream. When a particular CBR cell stream is *used* for MAC measurements, it must be as periodic as possible, to avoid measuring jitter introduced by the quasi-periodic cell stream.

3.7.4   Performance comparison

When studying the performance of different MAC protocols, two classes of performance measures can be distinguished:

- **MAC dependent measures** These measures are typical of and only relevant for a particular protocol. Two examples of MAC dependent measures are (a) the influence of the number of ONUs in a request block of the global-FIFO protocol on the end-to-end delay and (b) the impact on the end-to-end delay when a spacing unit is implemented in the McFred protocol. The analysis of these measures is too specific and falls outside the scope of this book. In [1] and [4] performance evaluations of MAC-dependent measures of respectively the BPP and the global-FIFO protocol can be found.

- **MAC independent measures** These measures are relevant to every MAC protocol and are therefore suitable for a performance comparison. An example of a MAC-independent measure is the delay which is due to the presence of a MAC protocol.

Since in performance studies of telecommunications systems quantiles of the distribution of a performance measure often play a key role, the performance evaluation is achieved by studying the stationary complementary distribution function (i.e. $Pr(X > x)$) of two important MAC-independent performance measures: the transfer delay and the CDV, introduced by the MAC protocols. Table 3.7.4 shows which line types are used in the figures of this chapter to present the performance of the MAC protocols introduced in Chapter 3.6. In Chapter 3.7.4.3

Table 3.7.4
Line types used in figures

Protocol Name	Corresponding line type
McFred	Dotted: .....
Global-FIFO	Solid: _____
BPP	Dashed: - - - -
DAA	Dashdot: - · - ·

Table 3.7.5
Parameters used

Protocol Name	Parameter values	
McFred	All scenarios:	$P_{\text{Max}} = 1 \rightarrow$ max. # cells in frame = 173
	Scenarios S1–S3:	$s_i = \lfloor (173 \cdot \sum PBR_i)/622.08 \cdot 10^6 \rfloor$
		$u_i = \lfloor 173/\# \text{ Active SMPCs} \rfloor$
	Scenario S4–S7:	$s_i = 0,\ u_i = \lfloor 173/\# \text{ Active SMPCs} \rfloor$
	Scenario S8:	$s_i = 0,\ u_i = \begin{cases} 8 & \text{SMPC } 0\ldots18 \\ 7 & \text{SMPC } 19 \end{cases}$
	Scenario S9, S10:	$s_i = 13,\ u_i = 17$
Global-FIFO	$\varepsilon = 0.8,\ T_{\text{Max}} = 43.21\mu s$	
BPP	8 RAUs in a frame of 144 slots	
DAA	$W_{\text{Min}} = 32,\ W_{\text{Max}} = 256$	

a short overview is given with respect to the overhead each MAC protocol requires to operate correctly.

Parameter settings   Each BAF MAC protocol has its own parameters, which have to be fixed before optimal use is guaranteed. The functionality of the parameters has been described in Chapter 3.6. Table 3.7.5 shows the parameter settings used to obtain the performance evaluation presented in this chapter. In this table the peak bit rate at $SMPC_i$ is abbreviated by PBR_i.

3.7.4.1   Comparison of transfer delays

The *Transfer Delay* of the BAF system is defined as the time difference between the sending of a cell at the T Interface and the receiving of this cell at the U Interface. The transfer delay of a cell consists of the following delay components.

1. A constant MAC-independent time (e.g. for encryption) $\approx$10 μs.
2. The time until a request can be sent.
3. The propagation delay of a request (50 μs).
4. Waiting time due to the permit distribution algorithm.
5. The permit propagation delay (50 μs).
6. The cell propagation delay (50 μs).

From the above definition it is clear that the transfer delay is a random variable depending on the protocol that controls access to the shared access link.

From Figures 3.7.1 to 3.7.3 it can be observed that when the BPP protocol controls access of the BAF System, the transfer delay is rather insensitive to the bit rate when homogeneous CBR traffic is considered (the dashed curve, which represents the transfer delay of the BPP protocol, shows the same behaviour in all figures). The transfer delay performance of the other protocols clearly depends on the bit rate of the CBR sources. For many low rate CBR connections the BPP protocol introduces the lowest transfer delay, whereas for high rate CBR connections the global-FIFO protocol and McFred protocol show the best performance.

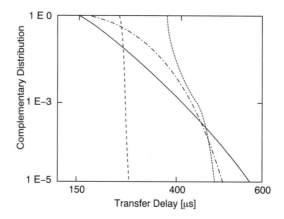

Figure 3.7.1
Scenario S1

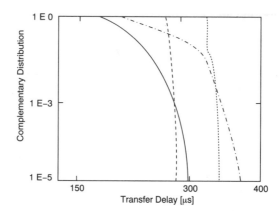

Figure 3.7.2
Scenario S2

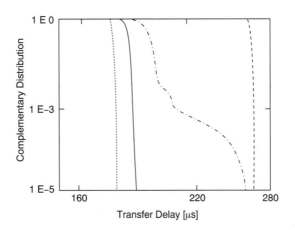

Figure 3.7.3
Scenario S3

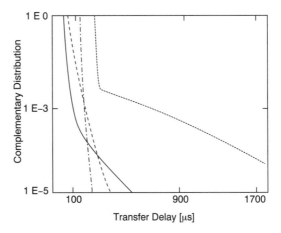

Figure 3.7.4
Scenario S4

Figures 3.7.4 to 3.7.7 show that for more bursty arrival patterns where temporarily overload conditions might occur, each MAC protocol individually causes more probability mass to be shifted to the tail of the distribution of the transfer delay.

The curves in Figure 3.7.8 which represent the transfer delay of the CBR source that uses ONU 19, are marked with an "X". Curves which are not marked with an "X" correspond to the transfer delays observed by those CBR sources which use one of the ONUs with number 0–18, which were loaded with heterogeneous traffic (see Table 3.7.3). From Figure 3.7.8 it can be observed that all MAC protocols offer fast access in the case of heterogeneous traffic and that when the BPP, the global-FIFO or the DAA protocol controls access the ATM network, there is a very small (and in some cases even negligible) difference in transfer

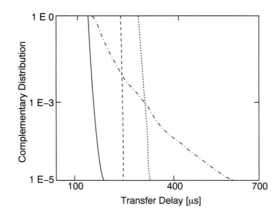

Figure 3.7.5
Scenario S5

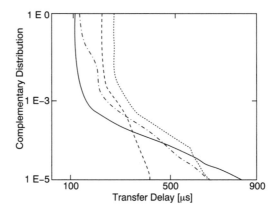

Figure 3.7.6
Scenario S6

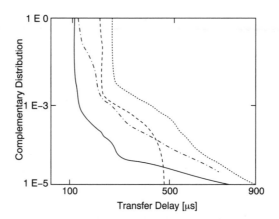

Figure 3.7.7
Scenario S7

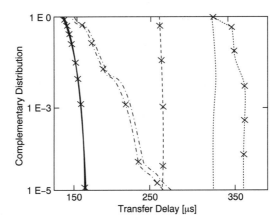

Figure 3.7.8
Scenario S8

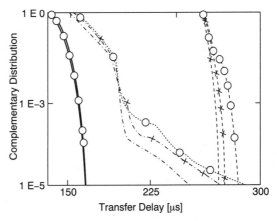

Figure 3.7.9
Scenario S9, S10

delay observed by the CBR connections which use a sole ONU and those which share the ONU with other connections. For the McFred protocol this difference is somewhat larger, whereas (contrary to DAA and global-FIFO) the McFred protocol offers a shorter transfer delay to those connections which share an ONU than to a connection which uses an isolated ONU.

In Figure 3.7.9 the curves representing scenarios S9 and S10 are marked with an "X" and an "O", respectively. The curves which are not marked are included for comparison and represent the performance in case where C-type traffic is replaced by an A3 source. From this figure it can be observed that the global-FIFO protocol shows a more stable performance, which indicates that the bundle spacing mechanism of this protocol provides a good mechanism for protecting the access network against unexpected bursty traffic.

3.7.4.2   *Comparison of the cell delay variation*

All MAC protocols have different features, introducing variation of delay. The most important reason why a BAF system causes cell delay variation (CDV) is that time must be gained in order to distribute permits efficiently. This is in all cases realised by making use of buffers. Variable queueing, VP shaping and multiplexing of VCs (all features of one or more of the described MAC protocols) cause variation in delay.

An accurate performance characterisation of CDV is a network performance parameter, known as 1-Point CDV. The 1-Point CDV describes the variation of the arrival times pattern with respect to the negotiated peak cell rate. It is measured by observing successive upstream cell arrivals at the U Interface and only considers cell clumping, i.e. the effect of cell inter-arrival distances which are shorter than T, which is the reciprocal of the peak cell rate. The characterisation of CDV by means of 1-point CDV was given in [6] and is recommended for CDV assessment by ITU-T. The 1-point CDV is defined as follows. Let a_k be the actual arrival time and c_k be reference arrival time of cell number k. The reference arrival time c_{k+1} is then computed as follows:

$$c_0 = a_0$$

$$c_{k+1} = a_k + T \quad \text{if } c_k < a_k$$

$$c_{k+1} = c_k + T \quad \text{if } c_k \geq a_k$$

The 1-point CDV y_k of cell number k is then defined as $y_k = c_k - a_k$.

Figures 3.7.10 to 3.7.12 show that the 1-point CDV performance of both the BPP and the McFred protocol is good and is also insensitive to the bit rate, when homogeneous CBR traffic is considered. The CDV performance of the global-FIFO and DAA protocol clearly depend

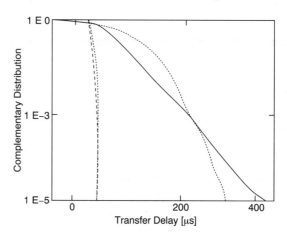

Figure 3.7.10
Scenario S1

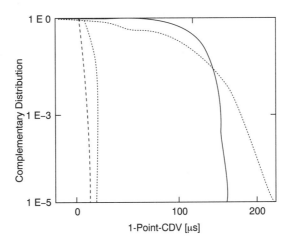

Figure 3.7.11
Scenario S2

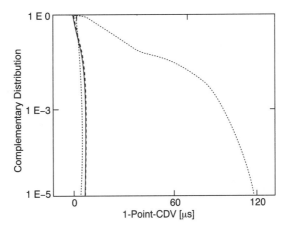

Figure 3.7.12
Scenario S3

on the bit rate of the offered CBR sources. For Scenario S3 the CDV introduced by the global-FIFO protocol tends to the results of McFred and BPP, whereas the 1-point CDV introduced by the DAA protocol remains high.

For more bursty arrival patterns where temporarily overload situations can occur, Figures 3.7.13 to 3.7.16 show that the spacing unit in the McFred protocol still provides low CDV performance independent of the bit rate. The 1-point CDV introduced by the BPP protocol clearly suffers more from temporarily overload conditions, but this protocol still provides better CDV performance than the global-FIFO and the DAA protocol.

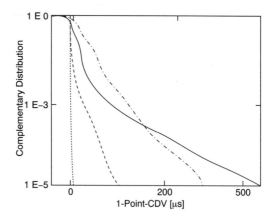

Figure 3.7.13
Scenario S4

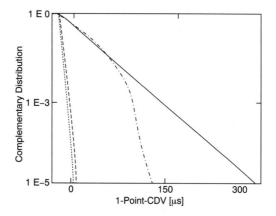

Figure 3.7.14
Scenario S5

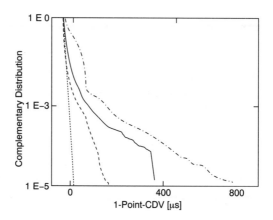

Figure 3.7.15
Scenario S6

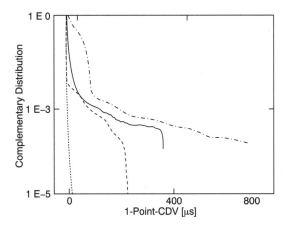

Figure 3.7.16
Scenario S7

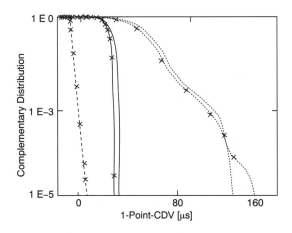

Figure 3.7.17
Scenario S8

Table 3.7.6
Overhead required

Protocol Name	Upstream	Downstream
McFred	2.833 bytes per ATM cell	0.833 bytes per ATM cell
Global-FIFO	3 bytes per ATM cell	3 bytes per ATM cell
BPP	8 slots in a frame of 144 slots	2 bytes per ATM cell
DAA	3 bytes per ATM cell	3 bytes per ATM cell

Figure 3.7.17 shows that for scenario S8 very little difference in CDV is observed between the CBR source which uses an isolated ONU and those CBR sources which share an ONU with other connections (the curves of the last one are marked with an "X"). The same CDV behaviour is observed for scenarios S9 and S10 (not shown).

3.7.4.3 Comparison of required overhead

In this subchapter a short overview is given of the extra upstream and downstream overhead required by each of the BAF MAC protocols. Since the McFred protocol is frame based, the number of overhead bytes per ATM cell strongly depends on the rate of the access link and on the number of assigned cells per frame.

REFERENCES

[1] Angelopoulos J D, *Time division sharing of tree PONs by ATM users: a method to control cell jitter*. Proc EUROPTO (SPIE/EOS) Conference on Broadband Strategies and Technologies for Local, Metropolitan and Optical Access Networks, Amsterdam, The Netherlands, March 1995

[2] Angelopoulos J D and Venieris I S, *A distributed FIFO spacer/multiplexer for access to tree APONs*, Proc SUPERCOMM/ICC Conference, New Orleans, Louisiana, May, 1994

[3] Blondia C *et al*, *Performance of Shared Medium Access Protocols for ATM Traffic Concentration*, European Trans Telecommunications, Special Issue on Teletraffic Research for B-ISDN in the RACE program **5**, (2), 219–26, 1994

[4] Garcià J, Blondia C, Casals O and Panken F, *The bundle-spacer: a cost effective alternative for traffic shaping in ATM networks*. Proc IFIP Conference on Local And Metropolitan Communication Systems, Kyoto, December 1994

[5] Casals O, Garcià J and Blondia C. *A medium access control protocol for an ATM access network*, Proc Fifth International Conference on Data Communication Systems and their Performance, High Speed Networks, North Carolina, Perros H and Viniotis Y (eds), October 1993

[6] CCITT Draft Recommendation I.35B *B-ISDN ATM Layer Cell transfer Performance*, 1992

[7] Parekha S and Warland J, *A quick simulation method for excessive backlogs in networks of queues*. IEEE Trans Automatic Control **34**, 54–6, 1989

[8] Sriram K and Whitt W, *Characterizing superposition arrival processes in packet multiplexers for voice and data*, IEEE J Selected Areas in Communications **4**, (6), 833–46, September 1986

[9] Shahabuddin P, Heidelberger P, Nicola V and Glynn P, *Fast simulation of availability in non-Markovian highly dependable systems*. Proc Twenty-Third International Symposium on Fault-Tolerant Computing 130–42, IEEE Computer Society Press, 1993

3.8 PRIVACY AND AUTHENTICITY IN PONS

3.8.1 Introduction

In this section the privacy and authenticity problems of an ATM PON with tree topology are studied. In conventional networking a frequently taken view is: privacy is a problem of the applications and should be solved by them. This approach is not applicable to the PON situation: PONs should become parts of the B-ISDN and as such they should be indistinguishable from other parts of the network from both an user's and an operator's point of view. The required homogeneous and ubiquitous service level can only be achieved by a service enhancement for the PON to meet the accepted standard of privacy and authenticity found with star-shaped networks. The situation is very much akin to the service level enhancement introduced by a CRC check in order to reduce the residual bit error rate on a link or a LAN. The measures have to be taken in the physical layer rather than in the application layer and therefore in our context the appropriate measures have to cope with data rates above 600 Mbit/s.

The system described in Chapter 3.2 is the reference system for the security analysis of this chapter. The chapter is organised as follows. First the threats are identified that would jeopardise the system, if no additional measures were to be taken. Then the solutions to the problems identified before are described.

3.8.2 Threat analysis

For a threat analysis it is useful to distinguish between upstream and downstream directions.

Upstream direction   Reflections of the upstream signal at the power splitters do not lead to readable information. Therefore eavesdropping on upstream information is not a problem. The real concerns are authentication and data integrity. A malicious user can impersonate an authorised subscriber in two ways, by a simulation attack or a substitution attack.

- A malicious user can send a cell with a VPI/VCI of another subscriber in his own allocated time slot — thereby generating new and unwanted information in another connection.

- A malicious user sends information (request blocks, ordinary ATM cells) at the same time as the authorised subscriber. As the receiver at the OLT has to cope with different power levels it might be possible to completely overwrite the correct information by sending at a higher power level. In contrast, transmission at the normal power level would typically produce a corrupted cell (loss of data integrity) which may or may not be detected by the application.

Downstream direction   As the OLT broadcasts the information to all ONUs, eavesdropping on information is possible. This applies to permits, OAM addresses and fast OAM commands, and the downstream ATM cells containing signalling information, user data, or OAM information.

Therefore the intruder can collect information about

- the amount of traffic generated and received by a subscriber — derived from permits and VPI addresses,
- the communication partners of a subscriber — derived from signalling information,
- the user data — derived from the payload of ordinary ATM cells,
- management policies and internal states of the system — derived from OAM cells and fast OAM commands.

3.8.3   Mechanisms to support privacy and authenticity in a PON

The threat analysis of the last section is an inventory against which the measures can be developed that give a PON with a tree topology virtually the same appearance as a star-type access network. Again, the upstream and downstream directions are considered separately.

Upstream direction   Message authentication and data integrity can be guaranteed with a very high probability, if redundant information is added which is dependent on the data and on a secret key of the originator. This redundant information is usually referred to as the "message authentication code" (MACO)[†] [11]. Message authentication, in general, is a procedure which, when established between two communicating partners, allows each partner to verify that received messages are genuine. In a PON environment with a tree topology, message authentication is only necessary in the upstream direction, because malicious users have no possibility of modifying downstream information. The benefit of the MACO has to be traded for the reduced transmission efficiency due to the additional redundant information.

In view of our threat analysis, modification of information is very likely to be detected by the application: non-interpretable sounds or pictures or — in the case of data transmissions — violation of CRCs will alarm the user. Overwriting of information can be circumvented

---

[†] In the literature dealing with cryptology Message Authentication Code is abbreviated by MAC

if a power control for all the involved transmitters is implemented, so that every transmitter is forced to send at a power level that leads to a received power at the root point within a certain specified range. This method excludes the possibility of a malicious user overwriting the upstream information of another user. If this was tried by some manipulation of the transmitter it would lead to an overmodulation of the receiver in the OLT.

Therefore the discussion is focused on the transmission attempts using a wrong VPI, i.e. a wrong identity. This problem can, however be circumvented by exploiting the knowledge represented in the MAC algorithm. Any upstream ATM cell is the result of a permit sent to the originator of that cell. Therefore the OLT can easily check whether the VPI of a received ATM cell matches the one expected according to the history of issued permits. This procedure is referred to as "origin checking". Origin checking has to be performed after a received cell has successfully passed the header-error check. As a result of the origin checking only authentic cells are forwarded by the OLT to the LEX. The LEX is then left with the problem that the identity claimed in the signalling information conveyed from the subscriber to the LEX might not be correct. But this attack is not typical for a tree network and has to be dealt with in a general attempt to enforce subscriber authentication before accessing network resources.

Downstream direction   To preclude unauthorised access to downstream information, encryption of this information is proposed. The idea of encryption is to disguise confidential information in such a way that its meaning is unintelligible to an unauthorised person. The information to be concealed is called the plaintext and the operation of disguising it is known as encryption. The encrypted message is called the ciphertext. The legitimate receiver of the message will be in possession of some secret knowledge, denoted as the key, that allows the inverse transformation from ciphertext back to plaintext. The process of recovering the plaintext from the ciphertext is known as decryption.

There are two important classes of encryption algorithms, symmetric and asymmetric. In a symmetric cryptosystem both communication partners have the same secret key k, an encryption procedure E to generate the ciphertext and a decryption procedure D to recover the original plaintext. A symmetric cryptosystem is outlined in Figure 3.8.1. The ciphertext c is generated from the plaintext p according to:

$$c = E_k(p)$$

and to recover the original plaintext p one performs the decryption operation based on the same key:

Figure 3.8.1
Symmetric cryptosystem

$$D_k(c) = D_k(E_k(p)) = p$$

In asymmetric cryptosystems the keys for encryption and decryption are different and it is computationally infeasible to derive one from another. Asymmetric cryptosystems are mainly used for key management and user authentication because their encryption speed only lies in the kbit/s range.

To meet a transmission speed in excess of 600 Mbit/s for the downstream direction, for the system described a stream cipher system is used. For encryption a pseudo-random sequence (PRS) generated by a pseudo-random generator (PRG) is added to the plaintext using bitwise modulo 2 addition. For deciphering the same algorithm has to be applied to the ciphertext. Such a stream cipher system is illustrated in Figure 3.8.2.

For $t \geq 0$, let $p(t)$ denote the binary plaintext stream and $k(t)$ the binary stream output of the pseudo-random generator denoted as the keystream. If $\oplus$ denotes the modulo 2 addition, then $c(t)$, the binary ciphertext, is defined by

$$c(t) = p(t) \oplus k(t).$$

The decryption process is given by

$$p(t) = c(t) \oplus k(t).$$

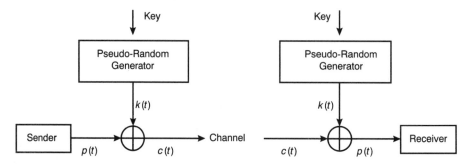

Figure 3.8.2
Stream cipher

The sender and receiver must possess the same keystream and this implies that they must be able to produce the same synchronised sequence of pseudo-random bits.

The circuitry needed to build up a PRG is based on linear feedback shift registers (LFSR), adders, and multipliers and can operate at very high data rates. The key represents the initial state of the PRG (e.g. intitial states or initial states and feedback coefficients of the LFSRs used). The key itself is a (pseudo-) random sequence. The PRG expands this (pseudo-) random sequence into a pseudo-random keystream. The encryption system has to withstand basic cryptanalytical attacks even under the assumption that a cryptanalyst has complete knowledge about the encryption system itself.

A set of general design criteria for keystream generators has evolved over time. Any secure keystream generator must satisfy this set of general design criteria. But a keystream generator which fulfils all design criteria need not be resistant to a generator specific cryptanalytic attack, if one exists (which cannot be excluded, even if none is found during the design of the generator). In this sense the design criteria form a set of necessary but not sufficient conditions for the security of a keystream generator. The most important criteria are the following.

- **The period of the pseudo-random sequence.** In general a PRG produces a periodic sequence. An infinite sequence (s_t), $t \in N$ is called periodic if there exists an integer $p > 0$ such that $s_p = s_{p+t}$ for all integers $t \in N$. If such an integer exists, then (s_t) is equal to $s_0 s_1 s_2 \ldots s_{p-1} s_0 s_1 \ldots$. The smallest value for p is called the period of the sequence. The period of the pseudo-random sequence should be much larger than the length of any message to be encrypted.

- **The linear complexity of the pseudo-random sequence.** Every finite or periodic sequence can be produced by a unique shortest LFSR. The length of this LFSR is said to be the linear complexity of the sequence. If the linear complexity is L and $2L$ consecutive bits of the sequence are known, the LFSR can completely be determined by the application of the Berlekamp–Massey algorithm [10]. In order to avoid stream reconstruction the linear complexity should be large.

- **The vulnerability to statistical attack of the pseudo-random sequence.** Ideally the sequence used in a stream cipher should comprise random data with the characteristics of white noise, i.e. given one section of the sequence, it would be impossible to predict any other section. Statistical tests to be performed on the output sequence of the keystream generator are described in [2], [3], and [6]. The basic idea of all these tests is to compare the results of the test performed on the pseudo-random sequence with those which one would get for real random sequences.

- **Confusion**. Every keystream bit must be a complex transformation of all or most of the key bits.

The work of [2] could be used as an introduction to cryptology. A good overview about keystream generators in general and their design can be found in [13]. A detailed discussion of linear feedback shift registers can be found in [8] and [9].

In the BAF system the shrinking generator [5], described below is implemented. This generator consists of two LFSRs; LFSR$_1$ and LFSR$_2$ of lengths m and n. This generator is illustrated in Figure 3.8.3.

LFSR$_1$ produces a pseudo-random sequence $A = a(t)$ of period 2^{m-1} and LFSR$_2$ produces a pseudo-random sequence $B = b(t)$ of period 2^{n-1}. The "shrunken" sequence $Z = z(t)$ is a subsequence of the sequence produced by LFSR$_1$ where the subsequence elements are chosen according to the positions of "1" bits in the sequence produced by LFSR$_2$. In other words, let $a_0, a_1, \ldots$ denote the first sequence and $b_0, b_1, \ldots$ the second one. The third sequence $z_0, z_1, \ldots$ includes those bits $a(t)$ for which the corresponding $b(t)$ is "1". Other bits from the first sequence are discarded.

Let m denote the length of LFSR$_1$ and n that of LFSR$_2$. If LFSR$_1$ and LFSR$_2$ have primitive feedback polynomials, i.e. the LFSR produces a pseudo-random sequence with maximal period $2^{\text{length of the LFSR}} - 1$, and $gcd(m, n)^\ddagger = 1$ then

- the shrunken sequence $z(t)$ has the period $(2^m - 1)2^{n-1}$ and
- for the linear complexity (LC) bounds can be found to be

$$m \cdot 2^{n-2} < LC \leq m \cdot 2^{n-1}$$

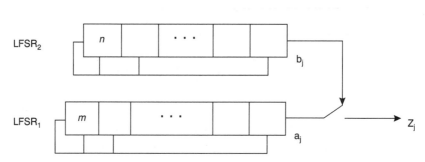

Figure 3.8.3
Principle of the Shrinking Generator

$\ddagger$ gcd: greatest common divisor

The way the shrinking generator is defined, bits are output at a rate that depends on the appearance of 1s in the output of $LFSR_2$. Therefore, this rate is on the average one bit for each two pulses of the clock driving the LFSRs. This problem has two aspects. One is the reduced output rate relative to the LFSR's speed, the other the irregularity of the output. These weaknesses must be overcome in a hardware implementation; on the other hand these weaknesses give most of the cryptographic strength to this construction!

One way of overcoming these weaknesses is to clock the LFSRs with the double required output rate of the shrinking generator, i.e. the rate of the data to be encrypted. Then with each clock pulse the shrinking generator can generate up to two output bits. Only one of them is required in the actual clock pulse. The spare bit is gathered in a buffer in order to compensate for sections of the encryption sequence when the shrinking generator has no output bit available. If there are no bits in the buffer the missing bit is filled with arbitrary values (e.g. alternating 0 and 1).

In the sequel the particular problems encountered in the encryption process as a result of interference from other processes such as ranging, routing or cell delineation, and as a result of synchronisation errors will be pinpointed. To minimise such risks it seems reasonable

- to do the ranging procedure without enabling the encryption procedure;
- not to encrypt the routing information, i.e. the VPI;
- to devise a cell delineation procedure which operates on the encrypted data.

The latter two issues and the synchronisation errors of the cryptosystem will now be considered in some more detail.

3.8.4   Encryption and routing

If the whole downstream ATM cell were encrypted with the key of the target ONU, every ONU would try to decrypt the cell in order to identify whether the VPI points to this ONU. However, this procedure is error prone because it might happen that besides

$$D_{k1}(E_{k1}(VPI_1, \ldots)) = VPI_1,$$

also

$$D_{k2}(E_{k1}(VPI_1, \ldots)) = VPI_2,$$

holds; in this case more than one ONU would receive a cell destined to only one ONU. Therefore it is proposed not to encrypt the VPI field.

From the viewpoint of the threat analysis, the transmission of plaintext VPIs might pose a problem, because the amount of received traffic per ONU is exposed. However, this problem can easily be circumvented by devising an algorithm by which the VPIs are only temporarily allocated to the ONUs. Owing to the VPI/VCI translation capabilities of the OLT, the LEX will not be aware of this behaviour. A change of the VPI of an ONU would be initiated by the OLT upon sending a corresponding (encrypted!) OAM cell to the respective ONU.

In the rest of this chapter it is assumed that

- a normal cell and
- an OAM cell

destined to a particular ONU can be distinguished by the VPI. Idle cells can also be identified from the VPI, but they will not be encrypted because they have no specific ONU as their destination.

3.8.5   Encryption and cell delineation

The standard procedure for cell delineation is to make use of the header error check (HEC) which will produce an all-zeros syndrome in the case where cell delineation is achieved (and transmission errors are negligible) [4]. To ensure that this procedure will also be applicable if cells are encrypted, the contents of the HEC field are recalculated after encryption and are modified accordingly.

3.8.6   Synchronisation of the ciphering/deciphering processes

At the OLT N ciphering processes exist, where N is the number of ONUs attached to the PON system (typically: $N = 32$). Each ONU runs only one deciphering process. The N ciphering processes at the OLT can be thought of as using the same PRG but started with a different key. A straightforward implementation at the OLT could make use of one PRG, which is then loaded with the key of the PRG of the ONU in question: see Figure 3.8.4.

When switching from one ONU to the next occurs, the state of the former ONU is again safeguarded in the memory. At the ONU side the PRG is clocked only upon the reception of a cell for this ONU. The keys of the PRGs at the OLT and at a particular ONU are given by a key exchanged via the safe upstream channel. The change of a key is described in the next subchapter.

The drawback of the approach described in Figure 3.8.4 is that the synchronism between an ONU and the OLT is dependent on the correct reception of cells by the ONU. Any misrouting or loss of cells will lead

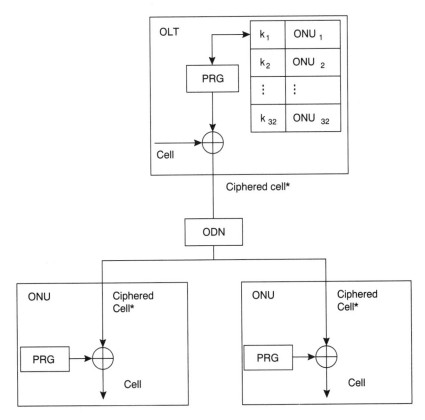

ciphered cell* = routing information is not encrypted

Figure 3.8.4
Downstream encryption using one PRG

to a loss of synchronisation in the encryption system. To avoid this, a sequence count on a per ONU basis would become necessary in order to cope with lost and misrouted cells. However, the corresponding control actions at the ONUs, which lost synchronism, would become rather involved. Therefore the approach depicted in Figure 3.8.5 is preferred.

Here the OLT has $N = 32$ PRGs which are clocked by the OLT's bit and cell clock. Thus for any cell to be sent downstream, each of the PRGs has to produce a PRS. The OLT selects only that PRS that corresponds to the addressee of the cell. Similarly, the PRGs at the ONUs make use of the same clocks derived from downstream information. Thus, as long as the cell delineation process works, synchronism at the encryption level is also guaranteed. This second alternative, described in Figure 3.8.5, requires a somewhat larger amount of hardware, but is a much simpler control scheme and therefore the preferred solution.

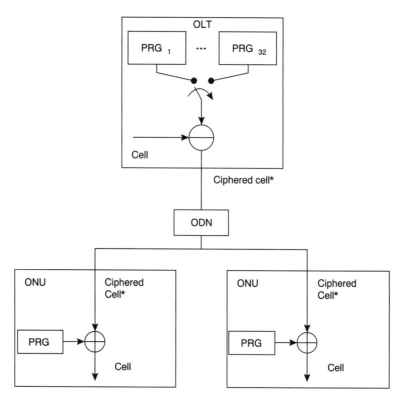

ciphered cell* = routing information is not encrypted

Figure 3.8.5
Downstream Encryption using several PRGs

3.8.7   Change of keys

If one assumes that some intruder was able to figure out the current
key (i.e. state) of one of the PRGs of the OLT, he would henceforth be
able to decipher all messages sent to the corresponding ONU. To limit
the extent to which the system can be exposed to the intruder a change
of key between the OLT and the ONU is carried out. The change-key
protocol will briefly be described.

The new key is generated by some (pseudo-) random process at the
ONU (see [1,7,12] for random processes). The initiative for a change of
keys is with the OLT; this is sensible, because the OLT which runs
many processes should preferably schedule its activities instead of
being interrupted by events. The OLT sends an OAM cell to the ONU
requesting a change-key. The Get response will then contain the new
key. The problem to be overcome is then, from which point in time
onwards the new key will be in use. The simplest answer could be that

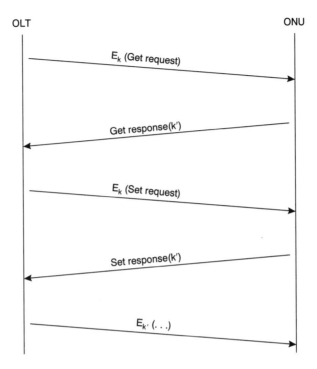

Figure 3.8.6
Change key protocol without fast OAM commands

the OLT issues an OAM cell requesting a switch to the new key. This is illustrated in Figure 3.8.6.

This figure corresponds to a four-way handshake. One of the problems with this solution is that the OLT might need to buffer cells for the corresponding ONU during the round-trip for the Set operation. In the system described here the aforementioned problem has been avoided by exploiting the concept of fast OAM commands. The use of the new key will then be initiated by the OLT by sending a fast OAM command piggybacked on the next downstream cell. This synchronisation may of course fail if the fast OAM command gets lost. There is a finite probability for this to happen even though the fast OAM address and command fields are protected by a CRC that can correct single and double errors. To guarantee a successful synchronisation the fast change-key command is followed by a slow OAM cell sent to the respective ONU.

A reasonable step of optimisation would be to piggyback the fast OAM command onto that OAM cell. The OAM cell would contain a Get status request. In case where synchronisation has been obtained, the respective Get response will contain the OK status. If synchronisation was not obtained the ONU would still be able to detect from the

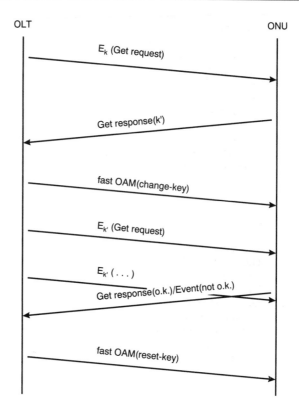

Figure 3.8.7
Change key protocol with fast OAM commands

non-encrypted VPI that it is receiving an OAM cell. But it will not be able to properly decode the message and therefore will issue an Event containing the not-OK-status. Thus the OLT is informed about the loss of synchronisation and will consequently issue a fast reset-key command to disable the encryption for that ONU. The change-key protocol with fast OAM commands is depicted in Figure 3.8.7.

Once the encryption system has been suspended and communication has been proven successful in this modus, the OLT can start the change-key protocol again. This is also the way the system will be initialised. Communication starts with the encryption system being suspended, and then keys are established for all ONUs on the basis of the change-key protocol.

REFERENCES

[1]   Agnew G B, *Random Sources for Cryptographic Systems*, in Advances in Cryptology — EUROCRYPT87, Lecture Notes in Computer Science 304, Springer 1988

[2]   Beker H and Piper F, *Cipher Systems: The Protection of Communications*, Northwood Books: London, 1982

[3]   Caelli W *et al*, *CRYPT–XS Statistical Package Manual for Stream Ciphers*, Queensland University of Technology, 1993

[4]   CCITT-Recommendation I. 432

[5]   Coppersmith D, Krawczyk H and Mansour Y, *The Shrinking Generator*, in Advances in Cryptology — CRYPTO '93, Lecture Notes in Computer Science 773, Springer, 1994

[6]   Dawson E P, *Design and Cryptanalysis of Symmetric Ciphers*, PhD Thesis, Queensland University of Technology, 1991

[7]   Fairfield R C *et al*, *An LSI Random Number Generator (RNG)*, in Advances in Cryptology — CRYPTO '84, Lecture Notes in Computer Science 196, Springer 1985

[8]   Golomb S W, *Shift Register Sequences*, Aegean Press, 1982

[9]   Lidl R and Niederreiter H, *Introduction to Finite Fields and their Applications*, Cambridge University Press: Cambridge, 1986

[10] Massey J L, *Shift-Register Synthesis and BCH Decoding*, IEEE Trans Inform Theory, **15**, 122–7, 1969

[11] Meyer C H and Matyas S M, *Cryptography: A New Dimension In Computer Data Security*, John Wiley & Sons, 1982

[12] Richter M, *Ein Rauschgenerator zur Gewinnung von quasi-idealen Zufallszahlen für die stochastische Simulation*, PhD Thesis, Rheinisch-Westfälische Technische Hochschule Aachen, Germany, 1992

[13] Rueppel R R, *Stream Ciphers* in Contemporary Cryptology, G.J. Simmons (ed), IEEE Press: New York, 1991

3.9 INTERACTION OF A PON SYSTEM WITH A B-ISDN LOCAL EXCHANGE

3.9.1 Introduction

This chapter covers some of the most critical issues related to the interconnection of the access network (AN) and the local exchange (LEX). The interface between the AN and the LEX is referred to as the V_B interface. A new definition of the V_B interface is given which reflects the B-ISDN/ATM environment and is general enough to cover a wide number of AN types. The novel AN–LEX interface is called the V_{B5} interface. The V_{B5} interface structure is described in detail both in terms of supported functions as well as in terms of physical realisation. For the latter an exhaustive list of application choices is given which apply to the employment of the V_{B5} interface in several cases, with respect to the number of LEXs and STM links.

To ensure the general applicability of the V_{B5} interface for any type of AN, it is necessary to carefully examine the allocation of functions to the different components of the environment in question, i.e. the user terminal equipment (TE), the AN and the LEX. In this sense, some particular functions which are considered to be critical for the ATM protocol operation are scrutinised. These include the identification of connections in complex combinations of T and V interfaces, the UPC operation for these connections, the identification of the service class of a connection. The purpose is to provide a list of potential solutions that enable the installation and operation of the AN without disturbing the transparency of the V_{B5} interface and the proper operation of ATM protocol functions in the LEX. For example, the AN internal need for addressing, service class indication and standardised connection identification is accommodated via an AN internal VPI value allocation algorithm, described in Chapter 3.2.5, which is kept transparent for the AN by virtue of VPI value translation tables located after the T and before the V_{B5} interfaces (see Figure 3.9.1).

After identifying the potential solutions regarding the location of the functions it is necessary to recommend how the information required for their application in the system might be obtained. More precisely, a fast control protocol for the exchange of information between the AN and the LEX should be developed. This chapter gives the general principles for such a protocol which has been called the local exchange access network interaction protocol (LAIP). LAIP was originally developed for ATM PONs as a fast control protocol running over the V_{B5} interface whose single-cell protocol data units (PDUs) are identified by a standardised VCI value. However, the LAIP messages and protocol procedures as specified here for each phase of connection establishment, connection characteristics modification and connection

Access to B-ISDN via Passive Optical Networks. Edited by U. Killat
© 1996 John Wiley & Sons Ltd

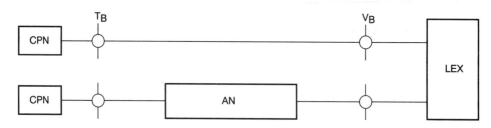

Figure 3.9.1
Simplified reference configuration. AN, Access Network; CPN, Customer Premises Network; LEX, Local Exchange

release have a broader scope that also covers other types of AN-like concentrator and multiplexer and can give solutions for service provision issues as in the case of best effort services; i.e., services with non-guaranteed bandwidth.

3.9.2 General description of the V$_B$ interface

There are two ways to connect a customer to a local exchange (LEX). A single customer may be connected to the LEX directly (e.g. via direct fibre) or via an access network (AN) that performs multiplexing and/or concentrating functions. The actual realisation mainly depends on the requirements that are made to the specific access regarding the demand on transport capacity. Large businesses will generally require more bandwidth than small business or residential customers. Concerning residential customers, this may change in the future if interactive applications like video on demand or teleworking become broadly available, and if they are accepted by the customers.

It is obvious that a customer extensively using broadband services (e.g. a large business customer) needs a high transport capacity and will therefore get access to the LEX via direct fibre. However, a customer only using the normal telephone service does not need a direct broadband access to the LEX. Instead he will be connected to the LEX via an AN that allows concentrating the traffic of many customers, thus enabling the sharing of infrastructure. This contributes to keeping the cost of the network access low, because it can be shared among many customers.

The access to the LEX is in any case (directly or via an AN) performed via the so-called V interface which is located on the customer side of the LEX (Figure 3.9.1). For the narrowband case this interface has already been standardised by ETSI [7,8]. For the B-ISDN/ATM case, i.e. the V$_B$ interface, the status of the work in the international standardisation community is still more or less at an initial stage. The standardisation of this interface is very important however, because only then is a network operator enabled to install and connect equipment from different manufacturers.

At the V$_B$ interface standard ATM cells are conveyed. The physical layer can be either SDH, PDH or pure ATM. According to the available standards [1,5,6] there is at present a preference to use SDH transmission systems. Some other characteristics such as transport bit rate, cell header format and encoding (UNI or NNI format), signalling procedures and support of distribution services depend on the actual access realisation and therefore cannot be generally defined. The relaying of timing information however must always be supported to allow correct bit (signal element) transmission and octet and cell delineation.

Concerning the handling of OAM flows, no special actions have to be taken in the case of a single customer directly connected to the LEX. The OAM information is transparently relayed over the V$_B$ interface to the ONU. The basic principles are defined in [4]. If an AN is installed between the LEX and the customer(s), the situation is a little different, because some of the OAM flows will be terminated in the AN. The standardisation of this issue is not yet settled, but it can be stated that it strongly depends on the actual realisation of the AN, e.g. whether the AN performs functions like UPC only on the VP level or also on the VC level.

Several different types of the V$_B$ interface are described in the standards [1,5,6], each of them for a different purpose and with specific characteristics. Figure 3.9.2 shows some possible configurations and the corresponding V$_B$ interface types on the LEX side. The V$_{B1}$ interface is used to connect a single broadband customer. Four different subtypes

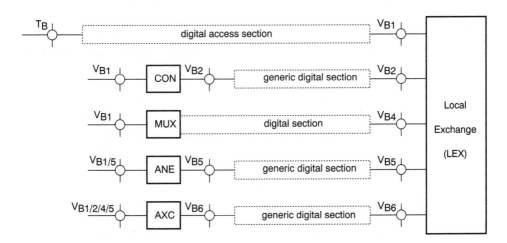

Figure 3.9.2
Possible V$_B$ interface configurations. ANE, Access Network element; AXC, ATM cross-connect; CON concentrator; MUX multiplexer

Table 3.9.1
Types of V_{B1} interface

interface	transport capacity	
	downstream	upstream
V_{B11}	149 760 kbit/s	149 760 kbit/s
V_{B12}	599 040 kbit/s	149 760 kbit/s
V_{B13}	599 040 kbit/s	599 040 kbit/s
V_{B14}	149 760 kbit/s	599 040 kbit/s

(V_{B11}, V_{B12}, V_{B13} and V_{B14}) have been defined, differing mainly in their transport capacities. As can be seen in Table 3.9.1, the subtypes V_{B11} and V_{B13} are symmetric and the subtypes V_{B12} and V_{B14} are asymmetric. The subtypes with asymmetric transport capacity can for example be used for provision of ATM distribution signals or interactive video requiring a broadband channel only in one direction. All V_{B1} types use the UNI coding scheme for the cell header as described in [3].

The other interface types (V_{B2}, V_{B4}, V_{B5} and V_{B6}) are used if an AN is installed between the LEX and the customer. These interface types differ in the access network element types that are supported. V_{B2} is a generic digital interface of a concentrator (CON) and supports signalling and controlling procedures between the access network element and the exchange. V_{B4} is a digital interface which does not appear explicitly at the output of the access network element itself, the multiplexer (MUX). This part of the subscriber access is strongly dependent on the individual realisation of the access network element. The V_{B5} interface supports a generic access network element (ANE) and will be considered in more detail in Chapter 3.9.3. V_{B6} is a generic digital interface of an ATM cross-connect (AXC). The network element is exclusively directed by management plane functions. Note that ANE stands for a single arbitrary network element in the access network area whereas AN represents the access network as a whole which may consist of a combination of ANE(s) and/or other elements such as a CON, MUX or AXC. For all these interfaces the ATM header layout is chosen according to the NNI format [3]. The reason is that within the ATM based AN, routing can easily be done on a VP basis. The use of the NNI header layout provides a larger addressing space because of its 12-bit VPI field.

3.9.3 Evolution to a universal multicustomer V_B interface

For the reason of cost reduction, the number of local exchanges is currently being reduced considerably. As a result each exchange serves a larger geographical area and more customers. This requires the installation of some kind of access network between the LEX and the customer. The V$_B$ interface finally has to provide the necessary functions to support multicustomer configurations. Sharing a part of the infrastructure is especially important for residential customers and small businesses. As was already mentioned in the previous section, the realisation of the access to the LEX via direct fibre is certainly not cost-effective for these customer groups.

The standards of ETSI and ITU currently describe configurations in the access network area including network elements with specific functions, such as concentrator (CON), multiplexer (MUX) or ATM cross-connect (AXC). The problem with these descriptions is that some realisations in the access network area do not allow identification of these separate/isolated network elements. In the case of ATM PONs for example, the functionality of CON, MUX and AXC may be distributed over the network element. As a result, the V$_B$ types V$_{B1x}$, V$_{B2}$, V$_{B4}$ and V$_{B6}$ are neither visible nor realised inside this configuration. Only an external V$_B$ interface can be identified.

If dynamic resource allocation is to be employed in the AN, the communication requirements between the LEX and the AN grow considerably. These requirements can no longer be met by the use of management protocols via the standardised management interfaces Q$_3$, X and M as in the case of static resource allocation by provisioning. This applies to all cases of dynamic procedures requiring a fast information exchange between the LEX and the AN. Therefore, a dedicated fast control protocol is needed between the LEX and the AN operating across the V$_B$ interface (see the following subchapters for more details).

For the above reasons the V$_B$ interface type V$_{B5}$ was introduced and is now being discussed in ETSI. It is used to connect arbitrary broadband access equipment to the LEX. To avoid possible confusion with existing functional network elements, the V$_{B5}$ supports any arbitrary access network element (ANE). The network elements AXC, CON and MUX support V$_{B1x}$ interfaces and partly also other V$_B$ types. For the case of ATM PONs a more general definition is necessary. T$_B$ interfaces also have to be supported, because the system may include the ONU. Therefore the ANE may support combinations of broadband customer access (V$_{B1x}$, direct T$_B$) and further ANEs (V$_{B5}$) to allow cascading. Figure 3.9.3 shows possible configurations including the V$_{B5}$ interface and an ANE. A specific implementation of an ANE does not have to support all possible configurations and external interfaces (T$_B$, V$_{B1x}$, V$_{B5}$).

Besides the functional architecture of the V$_B$ interface its physical realisation has a great impact on the characteristics of the whole access network. SDH transmission systems are now state-of-the-art and will therefore most likely be used for the physical structure of the V$_B$ interface, at least for the near future. So the remaining part of this subchapter

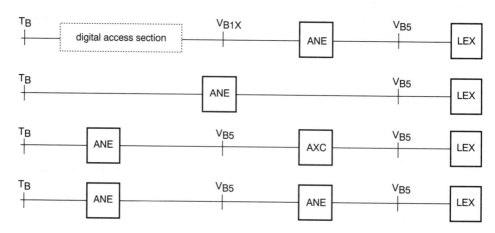

Figure 3.9.3
Configurations in the access network including a V_{B5} interface. ANE, Access Network element; AXC, ATM cross-connect; LEX, Local Exchange

focuses on possibilities but also on problems introduced by different realisations of the SDH interface ($1 \times STM - 1, 4 \times STM - 1, 1 \times STM - 4, 1 \times STM - 4c$).

The connection of an access network to the LEX may be based on one or more SDH link. STM-1 links with a gross bit rate of 155.52 Mbit/s and a transport capacity of 149.76 Mbit/s are used very often, because they do not require advanced technology. If more than one SDH link is used to increase the available transport capacity, for example $4 \times STM - 1$, the V_B interface will consist of several separate *physical* interfaces (four for the example of $4 \times STM - 1$). All physical interfaces that connect the access network to the same LEX represent one *logical* interface.

Having this in mind, configurations connecting one access network to more than one LEX via more than one logical interface are conceivable. This may be used to apply a kind of dual-homing of the AN to introduce redundancy and to realise some fault tolerance. But there is also another application. Currently the telecommunications market in many European countries is experiencing a major change. In the past each country had only one network operator offering telecommunications services. Meanwhile orders of the European Union and time schedules for the deregulation of the telecommunication market and privatisation of state owned operators exist. This process will lead to a situation with multiple network operators and service providers present in the same area. The V_B interface is identified as a possible boundary between these different operators and providers with respect to open network provisioning (ONP). As the AN is the most expensive part of the network, operators could agree to share the AN. In this case the access network would be for example connected via two separate logical V_B interfaces to two LEXs owned by different network operators.

If an access network is connected to the LEX via more than one *physical* interface, for example via $4 \times STM - 1$, the problem arises that data streams via different physical interfaces are independent of each other as far as the LEX is concerned. This means that the LEX handles standardised connections via $VPI = 0$, such as metasignalling or OAM flows, independently for each data stream. The same VPI/VCI values may be allocated to user connections for each data stream. The LEX does originally not know that all these data streams belong to the same logical interface, that they are transferred via the same access network and that they are thus *not* independent of each other. The access network element can solve the problem internally by translating the VPI values so that cells from different physical interfaces (at the LEX or at the customer side) are not mixed up (see Chapter 3.2.5). However, at the external interfaces V_B and T_B the connection related values are determined by the LEX and not the access network. Early implementations use a static resource allocation by provisioning. The range of available VPI/VCI values is split up between the physical V_B interfaces. So the VPI/VCI allocation becomes unambiguous. If it should be supported that connections entering the access network via different V_B interfaces are routed to the same T_B interface (connecting one T_B interface to the LEX via more than one physical interface at V_B), not only the number of V_B interfaces but additionally the number of T_B interfaces has to be taken into account for the division of VPI/VCI values. If a dynamic resource allocation is to be supported, a dedicated protocol (see the following sections) is necessary to ensure that for example the VPI allocation is unambiguous.

If there is more than one *logical* interface, i.e. the access network is connected to more than one LEX, the same difficulties arise. Each LEX handles standardised connections independently and allocates its own VPIs for user connections. If the multiple LEXs realise a multi-operator environment as described above, the agreement concerning the resource allocation even has to be made between different operators. It is obvious that a static solution is simpler to implement, but a method for the dynamic allocation is more flexible and is preferable with regard to a good utilisation of the resources.

Even if there is only one logical and one physical interface, problems may occur. If a higher multiplexing level with non-concatenated containers like $1 \times STM-4$ is used, the situation is the same as described above with multiple logical and/or physical interfaces. Each container in a non-concatenated synchronous transport module carries an independent data stream. These containers are multiplexed together for transmission but are still separate data streams. Thus, $1 \times$ STM-4 and $4 \times$ STM-1 are equal as far as the problems with resource allocation and standardised connections are concerned.

To avoid these problems an SDH interface structure has to be used with one logical interface, a single physical interface and concatenated containers, if a higher multiplexing level is necessary to increase the transport capacity. This is the case for realisations using $1 \times$ STM $- 1$ or $1 \times$ STM-4c.

3.9.4 Principles of controlling AN-LEX interaction over the V_B interface

The B-ISDN AN is located between the T_B and V_B interfaces and is intended also to support residential and small business customers. As explained in the previous sections, the limited budget and low bandwidth needs of the AN customers demand low-cost implementations that exploit network resources as much as possible. A way of supporting these objectives is an AN with no signalling capability, for example an AN operating on a virtual path (VP) basis, where the residential and small business customers will share a common medium. Another critical point is to ensure, through the standardisation of the V_B interface, a multi-operator AN with multi-vendor compatibility. This means that the AN internal topology and associated mechanisms should not affect the standardised T_B and V_B interface definition but should be kept as transparent as possible to the standard B-ISDN.

A high degree of transparency can be achieved at the expense of flexibility and efficient use of network resources. If bandwidth and VPI values are allocated statically to the AN VPCs, then the frequency of control message exchange between the AN and the LEX is low and can be accomplished by slow management procedures. The UPC operates on the VP level according to the predefined connection bandwidth but statistical multiplexing gain can be achieved only within a VPC. If the network operator wishes to use the full statistical multiplexing gain, then dynamic allocation of bandwidth to connections will be applied not only at the VC but also at the VP level. Under this scenario signalling-less network segments like the VP based AN which are not participating in the call establishment procedure remain isolated and unable to get information on the status of each VPC, which is also required for proper UPC operation. This information cannot be carried by management because of the very tight delay restrictions imposed on call/connection establishment. The problem becomes even more emphatic if one considers that the broadband signalling protocols will give the user the ability to change the bandwidth of an already established connection without passing through a release and re-establishment phase; e.g. when a user turns from a high to a lower quality service supported by the same connection.

As a direct consequence of the above requirements and to further optimise the use of the AN resources, a fast control interaction protocol, able to operate on a per call/connection basis between the AN and the LEX, has to be developed. This protocol has to offer an integrated solution to both current signalling-less VP based ANs operating via management procedures and the future VP based ANs allocating bandwidth to VPs via signalling. The services of the protocol should undertake the task of bridging the gap between static (the AN case) and dynamic resource allocation (the LEX case) by providing a means of dynamically updating both sides to the instantaneous state of each VPC. For simplicity and transparency reasons a master–slave mode of operation can be applied in the protocol with the LEX being the master. The messages to be exchanged should convey the relevant information accompanying a VPC establishment (VPC bandwidth and VPI value) or a VCC establishment that modifies the VPC bandwidth (VCC bandwidth).

3.9.5 A fast interaction protocol for APONs

In this section the Local Exchange Access Network Interaction Protocol (LAIP) is presented. The LAIP follows the concept established in the specification of the V5.1,2 interfaces intended for ISDN [7,8]. Its novel characteristics can be found in the ATM-related part of the LAIP which is centred around the use of VPIs/VCIs for identifying connections as well as its ability to allocate bandwidth dynamically so that statistical multiplexing is fully exploited. Other new issues appearing in the LAIP specification are related to the ability of the protocol to distinguish between the service classes supported by an integrated services AN so that different handling of cells at the AN is possible.

Each LAIP message is a single ATM cell. Since VCI standardisation is the way the ITU-T proceeds for identifying particular connections (e.g. metasignalling, general broadcast signalling etc.) a VCI value, i.e. VCI_L, at the V_B interface is employed for the LAIP connections. The VPI/VCI format of a LAIP cell becomes $[x, VCI_L]$ where x is the VPI value of the user information VPC in the AN or the V interface (for the VPI value translation functionality of the AN see Chapter 3.2.5). The AN should be able to access the VCI field of all cells and compare its value with the standardised VCI.

The fields of the LAIP messages are summarised in Table 3.9.2. Not all the fields of Table 3.9.2 are present in each LAIP message with the exception of the message type, declaring which LAIP message a cell

Table 3.9.2
The fields of the LAIP messages

Field	Length in octets	Values
Message Type (MT)	1	ALLOC
		ALLOC COMPLETE
		DEALLOC
		DEALLOC COMPLETE
		FAULT
		FAULT ACK
VP Reference Number (VPRN)	2	VPI value
ALARM	1	null
		BWind
		VPind
Fault Cause (FC)	1	possible fault causes
T Interface (T_{inf})	2	
V interface (V_{inf})	1	
VPI value at Tinf (VPI_t)	2	VPI value at T interface
VPI value at Vinf (VPI_v)	2	VPI value at V interface
Class	1	the service class
Bandwidth Requested (BW_r)	3	values in kbit/s
Cyclic Redundancy Check (CRC)	4	

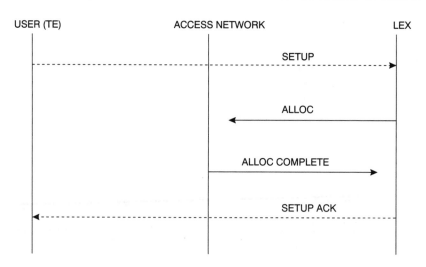

Figure 3.9.4
Message flow at the set-up phase

conveys, and the CRC field which is a 32-bit CRC following the specification of metasignalling protocol described in [11].

In Figure 3.9.4 the flow of messages between the AN and the LEX during the connection establishment phase is shown. The SETUP and SETUP ACK messages are standard signalling messages which pass transparently through the AN. Upon reception of the SETUP message the LEX issues an LAIP `ALLOC message to the AN in case the call/connection request is accepted by the LEX connection admission control. The ALLOC message contains the MT, VPRN, T_{inf}, VPI_t, V_{inf}, VPI_v, class, BW_r and CRC fields. The T_{inf} declares the identity of the T interface of the customer and is extracted from the SETUP message issued by the user. The V_{inf} is the identity of the particular V interface used by the LEX for this connection. The VPI_t and VPI_v are the VPI values allocated by the LEX to the T and V interface for the specific VPC. The class field declares the service class of the connection and the BW_r field is the bandwidth allocated to the specific VPC. With this information the AN is able to update the VPI translation tables (see Chapter 3.2.5) and the VP bandwidth use status required for the UPC operation, and to recognise the service class of the connection so that the corresponding cells are handled by the access control mechanism accordingly.

The VPRN field includes a common reference used by both the LEX and the AN for unique identification of a particular VPC. In principle a particular VPC represents a unique association among T_{inf}, VPI_t, V_{inf} and VPI_v as given by the following relation:

$$(T_{inf}, VPI_t, V_{inf}, VPI_v) \leftrightarrow \text{VPRN value} \qquad (9.3.1)$$

The same VPI_t value may be used in several T interfaces of a single AN. Hence the VPI_t cannot uniquely identify the VPC. The VPI_v can substitute the VPRN functionality only if a single V interface exists between the AN and the LEX and the AN is connected to only one LEX. In this case the following relation holds:

$$(T_{inf}, VPI_t) \leftrightarrow VPI_v \leftrightarrow \text{VPRN value} \qquad (9.3.2)$$

In the general case multiple V interfaces will be supported between the AN and the LEX and a T_{inf} can be connected to more than one V_{inf}. It then becomes possible for the same VPI_v value to be allocated to the different connections of a T_{inf} running over separate V_{inf}. In this case the VPI_v cannot uniquely identify the VPC. Instead of conveying the full set of four parameters indicated in relation (9.3.1) in every LAIP message, the VPRN field provides a means of indicating the particular VPC to which an LAIP message refers, by a single parameter. Recall that the LAIP operates in the form of single-cell messages and in this respect the more space available in the LAIP cell payload the greater the number of functions that can be introduced into the protocol.

Another possibility is that the same T interface is connected to several LEXs. In this case LEXs may allocate the same VPI_t value to different connections running over the same T interface. Also the same VPI_v value may be allocated to two V interfaces each one belonging to a different LEX. Again the use of the VPRN field becomes paramount but now relation (9.3.1) turns to (9.3.3):

$$(T_{inf}, VPI_t, V_{inf}, VPI_v, LEX_{id}) \leftrightarrow VPRN \text{ value} \qquad (9.3.3)$$

A problem arising here is the restriction in the number of VPRN values due to the addition of the LEX_{id} parameter. A potential solution to this problem is the adoption of a single LAIP entity (LPE) at the LEX level which plays the role of a coordinator for the several LEXs connected to the same AN. The single LPE is aware of the VPI values all LEXs allocate to the AN T_{inf}. Furthermore the single LPE is able to control the VPI_t value allocation procedure by defining a set of VPI values which an LEX can use for a particular T_{inf}. The set of these values is not necessarily static but it can be updated dynamically to protect rejection of a new connection owing to unavailability of VPI values at a particular T_{inf}. Each time a T_{inf} is physically connected to a new LEX the LAIP is activated. Its task is to reserve VPI values for the particular T interface and to inform the LEXs on the set of VPI values they can use for the connections running over the particular T interface.

The ALLOC COMPLETE message is sent from the AN to the LEX in response to the ALLOC message. It contains the MT, VPRN and ALARM fields. The ALARM field of the ALLOC COMPLETE declares in general whether the AN can accommodate a new VPC. Where there is no problem in the AN the ALARM field takes a "null" value. If a congestion situation is approaching because of the specific access control mechanism of the AN which in any case is kept transparent to the LEX (see Chapter 3.6) the ALARM field takes a "BW_{ind}" value. When the VPI values of the AN become exhausted the ALARM field signals the situation to the LEX via the "VPI_{ind}" value of the ALARM field. In any case the AN cannot reject a new call, because the LEX is the master, but can only notify the LEX about future unavailability of resources. Thus the "BW_{ind}" and "VPI_{ind}" alarms referring to a call connection request i should be taken into account when the $(i + 1)$th call/connection request arrives.

In Figure 3.9.5 the flow of messages between the AN and the LEX during the connection release phase is shown. The RELEASE and RELEASE ACK messages are standard signalling messages which pass transparently through the AN. Upon reception of the RELEASE message the LEX issues a DEALLOC message to the AN. The message contains the MT, VPRN, Class and BW_r fields. Again the fields of this message are used to update the VPI translation tables and the VP bandwidth use status required for the UPC operation. The DEALLOC COMPLETE message is sent from the AN to the LEX in response to the DEALLOC message and contains the MT and VPRN fields.

The FAULT message is sent from the AN to the LEX to indicate a fault condition within the AN. It contains the MT, FC and VPRN fields.

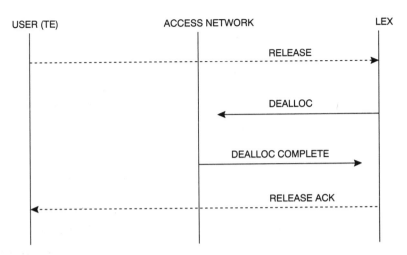

Figure 3.9.5
Message flow in the release phase

More than a single VPRN may be declared since it is possible that a fault condition in the AN affects multiple VPCs. The LEX replies with the FAULT ACK message in order to confirm to the AN that the fault condition will be taken into account. It contains the MT and VPRN fields.

3.9.6 Extension of LAIP for best effort service support in ANs

The main focus of the traffic engineering studies for ATM PONs is on investigating ways in which the ATM PON can accommodate services (like for example CBR services) with signalling capability and specific performance requirements that are guaranteed by the CAC algorithm of the LEX. For these services the ATM PON system relies on the assumption that the LEX accepts connections based on estimations that ensure that no congestion will occur in the network. It is then the responsibility of the ATM PON to provide mechanisms for preserving the connection contract. Services of this type are called guaranteed bandwidth (GBW) services. Apart from these services there are other types of services which present much less stringent requirements, for example the connectionless data. To keep the cost of such services low the obvious policy is to have system support only when there is spare bandwidth.

The main problem with these services is that there is no guarantee from the network side that this traffic will always find the required bandwidth or at least the necessary buffer space and therefore a higher than desired probability exists of rejection somewhere in the network in the case of congestion. This explains the term best effort services (BES) used to describe this service type. An exception here is when bandwidth is permanently allocated to the VPs supporting these connections. This case, however, is of no interest because it directly maps to the services described in the paragraph above and because it results in low utilisation of the system. Broadly speaking the network should always be able to accommodate a connection request even by allocating a bandwidth that is already used by a BES.

The AN as designed in the case of the BAF system does not provide excessive buffering capability for such services and in this respect the problem of loss becomes more emphatic. Moreover, current proposals introduce a throttling mechanism that undertakes the task of reducing the rate of such traffic when congestion is experienced. No such mechanisms have so far been investigated for the ATM PON.

In this section potential problems in the AN stemming from the BES concept are identified and mechanisms capable of supporting this type of service are proposed. These mechanisms are coupled to the LAIP which serves as a fast communication protocol between the LEX and the AN. The problem is formulated as follows. Assume that a request arrives at the LEX for a GBW service connection (either VPC or VCC). The LEX does not know how much of the BES bandwidth is actually used in the AN, but sees that the bandwidth α, where α is equal to the total bandwidth minus the bandwidth already allocated to GBW connections, is sufficient to accommodate the new request. So it establishes the new connection assuming that the BES cells that do not exceed a peak bandwidth value allocated to this type of traffic, and consequently are not rejected by the UPC, will be buffered somewhere in the network access components or will be kept in the user terminal equipment. The user begins to transmit GBW service traffic while the AN may experience a condition where the momentary rate of the injected traffic exceeds its capability. Hence the problem here is to define mechanisms in order (i) to preserve the QoS required by the GBW service connections in the AN and (ii) to avoid excessive loss of BES traffic cells.

Assume that the AN access mechanism is able to distinguish between service classes and to give priority to cells belonging to GBW services. Service class priority is realised in the permit allocation mechanism located in the OLT which distributes bandwidth selectively. This allows an easier distinction in the support of GBW services as opposed to BES but does not automatically solve the problem of extensive losses of BES cells. The services of a throttling mechanism are needed if generous buffering capability is not provided.

If a service transparent access scheme is assumed the OLT will allocate permits based on the total number of cells residing in the ONU. To avoid allocation of bandwidth or permits to BES traffic while GBW service traffic remains unserved the ONU should maintain at least two separate queues, one for each type of service. The particular queue in which an incoming cell has to be placed is recognised by the VPI_t

of the cell which implicitly declares the service type of the cell. This is achieved by the LAIP service class parameter (see Chapter 3.9.5) which is linked by the AN to the internal VPI value (see Chapter 3.2.5) used to establish a unique association among the VPI_t, T_{inf}, VPI_v and V_{inf} parameters. The maintenance of two separate queues in each ONU partially solves the problem of GBW service performance. The reason is that at the ONU level, the permit allocation algorithm will always give priority to GBW service cells without preserving this priority in the entire AN. This means that there is always the possibility that BES cells are forwarded or that permits are given to an ONU with BES traffic while other GBW service cells residing in other ONUs do not receive enough permits. (This is exactly the same situation faced in the DQDB global queue [10] which although fair for nodes, cannot distinguish between services). Again this problem can be solved by using a throttling mechanism that could operate at the source (user terminal), based on the congestion perception which the OLT obtains from the rate of requests.

Independently of whether the access control mechanism executes priorities or not, the requirement for guaranteeing a low probability of loss for BES traffic cells still exists. A way to accommodate this requirement is the introduction of a throttling mechanism bringing back feedback information to the users to regulate the BES information flow. Without this mechanism the users remain unaware of the congestion conditions and continue to transmit at their current rate until eventually BES cells are rejected in the ONU buffers. The most promising throttling mechanism currently proposed for ATM LANs is the so called backward explicit congestion notification (BECN) algorithm [9]. Cells conveying a congestion notification indication are issued from the ATM switch to the terminals to signal that losses are imminent and that the user should reduce or even stop the transmission of BES traffic. For the case of the ATMPON considered here the following alternatives emerge regarding the location of the throttling mechanism.

ONU   The BECN cell is issued by the ONU when the BES queue exceeds a certain threshold. Actually this means that there are a lot of GBW cells and that there are not enough permits to satisfy both types of service. The drawback of this solution is the implementation complexity of the ONU.

OLT   The BECN cell is issued by the OLT and passes transparently through the ONUs to the user terminal. Congestion is detected in the OLT by using the knowledge of the permit allocation algorithm. There the discrepancy between requests and permits for BES in the case of distinct priority cells is quantified for each ONU, allowing an estimation of the growth of the BES. When no priorities are realised the same restriction on BES will be initiated on the basis of total requests exceeding a threshold. The advantage of this solution is that the throttling mechanism is realised in the OLT without adding any significant implementation complexity. Note that the OLT is already complex owing to the execution of the permit allocation algorithm. The only concern is the reaction delay of the throttling mechanism; i.e., a congested ONU declaring its state after waiting for a transmission permit.

Extension of the LAIP   The BECN cell is issued by the LEX and passes transparently through the AN to the user terminal. To support this mechanism the LAIP services are required for notifying the LEX about the situation in the ONUs. This is done in the same way as in the OLT case; i.e., by using the knowledge of the permit allocation algorithm functional block or of congestion indication being received by the ONUs. In such situations the OLT issues a LAIP message to the LEX initiating a corresponding action by the LEX; that is, the LEX generates a BECN cell which will be forwarded transparently through the AN to the user terminals feeding the congested ONU(s) BES queue. The LAIP message initiated by the OLT must indicate the specific T_{inf}/ONU where congestion is experienced. This can be done either by putting the T_{inf} identity in the LAIP message payload or by using the VPRN field. In the last case the VPRN value should be associated with the T identity or in other words it should correspond to the VPI(s) of the T_{inf} supporting BES traffic. The LAIP cell may use any of these VPIs in its header.

REFERENCES

[1]   RACE Common Functional Specification E320, Issue E, *Interface at the V_B Reference Point*, 1994
[2]   ITU-T Recommendation I.311 *B-ISDN General Network Aspects*, 1992
[3]   ITU-T Recommendation I.361 *B-ISDN ATM layer specification*, 1992
[4]   ITU-T Recommendation I.610 *B-ISDN OAM Principles and Functions*, 1992
[5]   ITU-T Draft rec. Q.51A *Interfaces on B-ISDN Network Nodes*, 1993
[6]   ETSI ETR RTR/SPS-3025 *Signalling Protocols and Switching — B-ISDN Switching, Exchange and Cross Connect Functions and Performance Requirements — Enhanced Version*, 1993
[7]   ETSI DE/SPS-3003.1, version 8A *V5.1 interface for the support of Access Network (AN)*, 1992
[8]   ETSI DE/SPS-3003.2, version 7A *V5.2 interface for the support of Access Network (AN)*, 1993
[9]   Newman P, *Traffic management for ATM local area networks*, IEEE Comm Mag, **32**, (8), 44–50, 1994
[10]   IEEE 802.6, IEEE Standards for Local and Metropolitan Area Networks *Distributed Queue Dual Bus (DQDB) Subnetwork of a Metropolitan Area Network*, 1991
[11]   ITU-T Recommendation Q.2120 B-ISDN Meta-signalling, 1994

3.10 MANAGEMENT OF AN ATM PON

3.10.1 Introduction

The administration and management of the recently developed telecommunications networks is a challenge of its own. A series of sophisticated management systems has been developed to cover all apparent problems. The missing compatibility between these systems led to the introduction of the telecommunications management network (TMN) concept. The TMN is an independent information processing system. It enables network operators to supervise and control components of their telecommunication networks via standardised interfaces and protocols. The TMN is a separate network that interfaces to a telecommunications network at several points to receive information from it and to control its operations. This TMN often uses different parts of the telecommunications network for its own communication. In Figure 3.10.1, the generic TMN architecture applied to the telecommunications network is shown.

In meeting these demands, a network operator will try to develop more efficient and flexible operations systems (OS) that conform to standard recommendations for telecommunications network management. Standardisation of management systems is currently being studied by several bodies such as ITU-T, ETSI, and EURESCOM, and today is still an open issue.

The access network described in previous chapters, based on shared passive optical fibre and ATM protocols, encloses specific communication functions and procedures (MAC, ranging, bandwidth allocation, etc.) that should be managed by the operations system. In this chapter only the management view of these procedures and functions will be discussed.

This chapter is composed of two main parts: the first part introduces the management requirements of an access network based on a passive optical network (PON), and the second part analyses in depth one of the management approaches used to define and develop the operations systems for the access network under study.

3.10.2 PON management requirements

The management requirements of the AN are evaluated here by considering different management areas, each representing a different viewpoint of the management. As defined in the corresponding ITU-T recommendations [5], there are essentially five areas: performance, fault, configuration, security and accounting, which will also be used here to illustrate the management requirements of an ATM PON.

Access to B-ISDN via Passive Optical Networks. Edited by U. Killat
© 1996 John Wiley & Sons Ltd

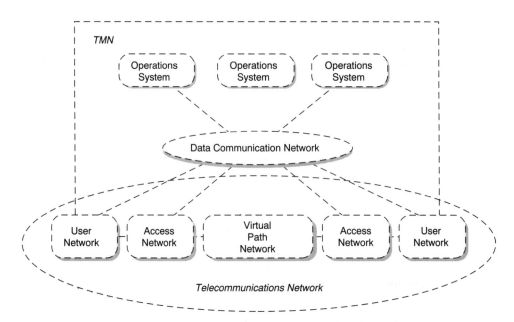

Figure 3.10.1
Relationships between a TMN and a telecommunications network

3.10.2.1   Performance management

The monitoring of service characteristics provided to the customer by the AN is done by the performance manager. Although the performance manager should monitor all the ATM traffic characteristics specified in ITU recommendations, this area is mainly concerned with monitoring the cell loss rate (CLR) on a per-VPC basis. Cell loss, in the upstream direction, can be due to overflow of internal queues and also to the conversion of the transmission rate from 622 Mbits/s to 155 Mbits/s in the OLT. Performance management takes action to correct loss of performance by interacting with configuration management to adjust, for example, the parameters of the MAC controller to guarantee users' access to the agreed bandwidth of their VPCs. It will also send alarms to the fault manager if the performance is degraded beyond a defined limit.

3.10.2.2   Fault management

The fault manager provides the alarm/failure detection, testing, isolation and recovery measurements for the different network elements (NEs). The fault messages (events, alarms and fault notifications) will be carried in operation and maintenance (OAM) cells transmitted from the NEs to the OS via the Q3 management interface. As the PON uses passive components the fault manager is unable to manage and supervise the transmission infrastructure composed of splitters, combiners and fibres. Therefore, the fault manager will only receive fault information from the access network NE, related to internal faults (loss of local power and environmental events), transmission faults (loss of signal, loss of bit timing, loss of upstream/downstream clock, upstream collisions, cell header error, loss of frame alignment). It will also receive faults related to the medium arising from its degraded performance or due to traffic interruption (error performance above a given threshold or excessive error rate).

The fault manager is also responsible for communicating relevant information to other interested management functions included in the rest of the functional areas. The purpose of communicating this information is to update the status of the AN.

3.10.2.3   Configuration management

According to the ISO description, the configuration functional area includes the set of functions that exercises control over the system configuration and components within a system. This set of functions will manage the installation or shut-down of both, hardware and software AN components, and should also manage the name relationship between components installed. The configuration manager is not only involved in the static installation procedures, but also in the dynamic reconfiguration of the components installed along the distributed environment which is represented by the AN.

Within the framework of the configuration manager, for an AN based on a PON infrastructure the main configuration requirements are:

- The management of static and dynamic ranging mechanisms. These procedures should include the delay time calculations and the interaction with the fault manager in order to perform corrective actions.
- The management of VPI values. These functions include routing and addressing inside the AN, access priority and policing at the edge of the AN.

3.10.2.4   Security management

The access network based on a PON with tree topology is a shared environment and requires some privacy and security functions to protect the customers from malicious users. Privacy involves protection of the information sent or received by the customer from being accessed by anybody, and security means the protection of the information from being modified by another user. The security manager is responsible for the monitoring and managing of exceptional states related to the protection and security of the information flowing across the access network.

Another issue related to the security of management is introduced by the fact that the AN will be owned/operated by one or more operator companies. In a shared ownership/operation scenario the management information will be exchanged between operators through an X interface and although there will be some security functions at this interface, security management will be responsible for authorising other operations system (OSs) to access management information.

Security management interacts with the rest of the functional areas. There should be security functions residing in the performance, accounting, fault and configuration areas, but it is in the security functional area where the generic security mechanisms applied to the access network are defined.

3.10.2.5   Accounting management

The functionality of the accounting area is closely related to performance management, as the data on which the accounting is based are obtained from the performance monitors existing in the different network elements (NEs) of the AN. The performance information collected by the AN NEs will be sent over the X management interface from AN-OS to the B-ISDN-OS. The B-ISDN-OS will use the performance data received from the AN-OS to calculate the real data carried over the user connection and will bill the customer accordingly.

3.10.3   TMN approach

The development of a management system for an access network can be based on either the ITU standard CMIP model, or the industry standard SNMP model. These two standards lead to a management system with different attributes and will briefly be described here.

The SNMP (simple network management protocol) model defines a simple method for the structure of managed information (SMI). Management data is stored as a hierarchical tree in the database (management information base (MIB)), without the inheritance concept. However, the main disadvantage is that the SNMP-MIB is determined at its design time and remains static during the lifetime of the system.

In contrast, the CMIP model is a more dynamic approach using primitives such as 'create/delete/retrieve/set' to instantaneously modify the structure of the database. This provides a far more flexible and efficient management of data over the AN. However, it has been criticised for being complex and too rich in functionality compared with SNMP, and so it will lead to a longer development period. Even after considering this disadvantage, it can be seen that the TMN approach is being adopted by more and more developers, and as more 'off-the-shelf' software becomes available, the production life-cycle will be shortened. This reusability property is made possible by using advanced information processing techniques such as object-oriented software engineering (OOSE). In the OOSE approach, the resources of the AN are defined as objects that need to be managed and controlled by the OS. As more generic network resources are defined as managed objects (MOs), these can be reused and developers will see a reduction in the production lifetime cycle.

Selection of SNMP implies a relatively lean stack, which provides the principal advantage of fast execution and the consumption of a relatively small amount of memory, but the static MIB provided and the missing inheritance concept will be reasons enough to stick to the TMN solution.

Adoption of the TMN approach to a broadband access network, based on PON and ATM concepts, implies the network architecture shown in Figure 3.10.2. In a deregulated telecommunications environment, three different ownership domains for the AN can be defined: the customer domain containing the customer premises equipments (CPE), the access network domain and the public network domain, including the local control related functions (LCRF). In order to manage the telecommunications resources included in these domains, at least one operations system per ownership domain should be developed. In that case, the ITU-T recommendations state that X-type management interfaces should exist between these ownership domains and inside the TMN domains Q3 type management interfaces should exist.

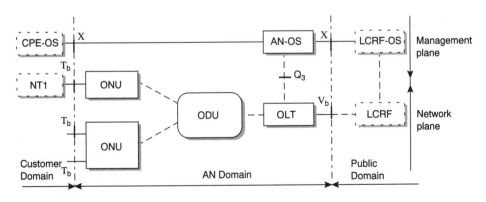

Figure 3.10.2
Management architecture

It is also possible that the OLT might support a non-standard interface (called an M interface) instead of a Q3 interface, as shown in Figure 3.2.2.1, in order to reduce the complexity of the NE. As the M interface is not a TMN standard, an additional adaptation function called the Q-adaptation function (QAF) needs to be implemented to translate nonstandard TMN data to a format accepted by a TMN interface. A description of the difference in these interfaces is given in Chapter 3.10.4 (TMN physical architecture).

The management architecture defines one Q3 interface for the access network between the operations system (OS) and the OLT equipment. The management information between ONU and the OLT is carried over the AN network itself by slow OAM and fast OAM commands.

Fast OAM commands are used to transmit the management information with tight time constraints from the OLT to the ONU; they are encoded in the preamble of an arbitrary downstream BAF cell and are protected by a forward error control mechanism. By using fast OAM commands, delays in issuing control information can be avoided.

Slow OAM cells, which carry the control messages, have the same header format as user cells at the network-network interface (NNI). These cells are inserted by the OLT and each ONU into downstream and upstream traffic flows, respectively. There will be a default non-user virtual path connection (VPC) set up between the OLT and each ONU for exchanging these internal OAM cells. A theoretical maximum of about 32 Kbit/s of bandwidth, in both directions of data transfer may be allocated to each of these internal VPCs. It is important to notice that the transport of slow OAM cells over the network is not guaranteed. The management protocol should specify that each request (sent by the manager) has a corresponding response (sent by the agent). So the NEs will use a simple retransmission protocol to deal with lost control messages.

As recommended in the TMN standards the following subchapters describe the three different viewpoints of the TMN architecture: functional architecture, information architecture and physical architecture. Each subchapter introduces the concepts used in the viewpoint definition; for the information architecture the respective concepts applied to the access network under study are described in quite some detail.

3.10.3.1 TMN functional architecture

The functional architecture describes the distribution of functionality within the TMN. This functionality is represented by function blocks. These blocks are contained in building blocks through which a customised TMN can be implemented. Figure 3.10.3 illustrates the function blocks, as they are proposed by ITU-T recommendations, and it also indicates the functions that are directly involved in

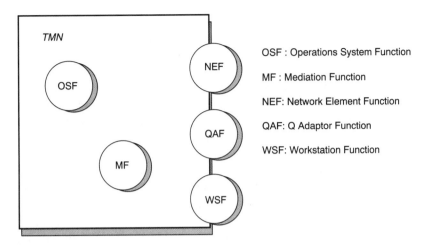

Figure 3.10.3
Function blocks in the TMN

TMN management. Note that some of the function blocks are shown as only partly belonging to the TMN. The functionality provided outside the TMN is not subject to standardisation. The following is a description of the different function blocks defined in TMN recommendation [5].

The operations system function processes information related to telecommunications management for the purpose of monitoring, coordinating and controlling telecommunications functions. It represents the main functionality needed to manage the telecommunication network and interacts with the other functional blocks involved in the network management.

The mediation function acts on information passing between an OSF and NEFs (or QAFs) to ensure that the information conforms to the expectations of the function blocks attached to the MF. This may be necessary as the scopes of the information supported by different communicating function blocks at the same reference point can differ. Mediation function blocks may store, adapt, filter, threshold and condense information.

The network element function (NEF) communicates with the TMN for the purpose of being monitored and controlled. It includes the telecommunications functions that are subject to management. These functions are not part of the TMN but are represented to the TMN by the NEF. The part of the NEF that provides this representation is part of the TMN itself, while the telecommunications functions themselves are external to TMN.

The Q-adaptor function (QAF) block is used to connect non-TMN entities which are NEF-like and OS-like to the TMN. The responsibility of the QAF is to translate between a TMN reference point and a non-TMN (e.g., proprietary) reference point and hence this latter activity is shown outside the TMN.

The workstation function (WSF) provides the means to interpret TMN information for the management information user. This block manages user terminals to display information, to transfer user commands, etc.

In order to delineate management function blocks and to define the information that flows between them, the concept of a reference point has been introduced. It simply identifies the information passing between function blocks and its syntax. Three types of TMN reference point have to be distinguished:

- The *Q reference point* governs the information exchange between function blocks defined in the same ownership domain with common management information supported by these blocks. This type of reference point is always defined between function blocks belonging to the same TMN.

- The *X reference point* is located between OSF Blocks in different TMNs. Entities placed beyond an X reference point may or may not be part of an actual TMN.

- The *F reference point* describes the interface necessary for the communication of a WSF block with any other TMN internal function block.

3.10.3.2 *TMN information architecture*

This architecture employs an object-oriented approach for transaction-oriented information exchanges. In this section the concepts of information modelling are presented and are then applied to an access network by defining its corresponding information model.

A management information model represents an abstraction of the management aspects of network resources and the related support management activities. It is composed of managed objects which are resources for the purpose of management. A managed object is specified by operations, attributes and notifications visible to management, whereas the internal functioning of the resource is not visible to management. This leads to the encapsulation of all relevant data within an MO.

Specifications of a managed object class are made by using templates defined in [19]. The definition of a managed object consists of

- the position of the managed object class in the inheritance hierarchy,

- a collection of mandatory packages of attributes, operations, notifications and behaviour,

- a collection of conditional packages of attributes, operations, notifications and behaviour. The condition under which each package will be present must be also specified.
- Within the package structure:
  - — Attributes visible at the managed object boundary,
  - — Operations which can be applied to the managed object,
  - — Notifications that can be emitted by the managed object,
  - — The behaviour exhibited by the managed object.

Other templates (name bindings) specify the possible superior objects for instances of a given managed object class, as well as the attribute used for naming. The collection of name binding templates will join the naming tree, where the hierarchy is organised on the basis of the containment relationship.

Another major concept associated with the MO is the inheritance between classes. Inheritance is the conceptual mechanism by which attributes, notifications, operations and types of behaviour are acquired by a subclass from its superclass. Multiple inheritance is the ability of a subclass to be specialised from more than one superclass. Specialisation by deleting any of the superclass characteristics is not allowed.

Network management is a distributed application since the environment being managed is distributed. This involves the exchange of management information between management processes to monitor and control the various networking resources. For every management association the management processes will take on one of the two following roles.

- Manager role, assumed by the part of the distributed application that issues management operation directives and receives notifications.
- Agent role, represented by the part of the application process that manages the associated managed objects. The role of the agent will be to respond to directives issued by a manager. It will also present to the manager a view of these objects and emit notifications reflecting the behaviour of these objects.

For example, these roles can be mapped onto an OS function setting up a VPC. Here, the OS (acting as manager) instructs a VPC to be set up between the OLT and ONU, and assigns a specific ID and attributes. Figure 3.10.4 shows the interaction between manager, agent and objects. It should be noted that this model is only an abstraction for the purpose of management. The internal implementation of an agent is not a subject for standardisation. The only standardised feature of the agent is the interface it offers when it is considered as a managed system.

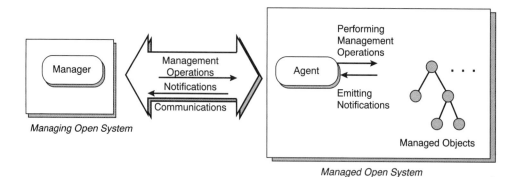

Figure 3.10.4
Management view following TMN concepts

Access network information model   The aim of the AN information model is to define the information exchanged at the management interface located between the OLT and the OS. Such information is modelled using design principles outlined in the management information model [19]. Resources are modelled as objects, and the management view of a resource is called a managed object (MO). Additional objects, called support managed objects, are defined to support the functions of managing the AN.

In order to define a TMN standard operations system (OS) for the AN and using the reusability concept introduced by ITU-T, the support objects are imported from the standard MO libraries, defined by ISO (generic MOs), ITU-T (generic network MOs) and ETSI (ATM, SDH and PDH MOs). The AN management capabilities are located at three levels of abstraction: network element view, network view and service view. As defined in [7], the network element view is concerned with the information that is required to manage a given network element. The network view is concerned with the information representing the network, both physically and logically. It specifies how network element entities are related, topographically interconnected and configured to provide the connectivity. Finally, the service view is concerned with how network level aspects are utilised to provide a network service.

The possibility of reuse with a subsequent increase in productivity and quality has been recognised by ITU-T and ISO who have developed generic management information models. For example, areas such as synchronous digital hierarchy can make use of these generic MOs. Within the PON access network, some of these standard objects may be reused. These are the main recommendations that are applicable: ITU-T M.3100 [7]; ITU-T X.721 (ISO/IEC 10165-2) [8]; ITU-T G.774 [3]; ETS DE/NA5-2210 [2].

The MOs included in the AN information model are structured in fragments. These fragments make up the complete information model. A fragment encloses all the MOs dealing with a particular subject, but object classes of each fragment will be usable in various models. The purpose of defining fragments is only to have a document that is easier to read by grouping a limited number of object class definitions. The AN object classes are grouped into 14 different fragments:

1. The generic fragment includes the MOs related to the alarm handling area, generic classes used for naming (system) and management support objects (powerSupply). Their inheritance and entity-relationship diagrams can be found in ITU-T and ETSI recommendations [7,10,18] and they will be useful in the definition of the AN information model.

2. The AN specific generic fragment includes generic MOs defined for different purposes in the access network, which represent the specific AN resources to be managed, such as specific pieces of software.

3. The generic termination point fragment groups those MOs that are defined in [7] which represent the beginning and/or the end-points of a generic link connection or transport entities such as trails and connections. In this fragment the generic concepts, defined in [7], of trail termination point (TTP) and connection termination point (CTP) are used.

4. The ATM termination point fragment includes MOs defined in [2] which represent the beginning and end-points of an ATM path as specific termination points.

5. The SDH termination point fragment includes MOs defined in [3] which represent the beginning and end-points of an SDH connection as a CTP.

6. The PDH termination point fragment groups MOs defined in [11] and will be used to represent the management capabilities of the AN PDH access points.

7. The AN specific termination point fragment includes MOs representing AN internal termination points, between OLT and ONU. They have been specifically defined for the PON based access network.

8. The cross-connection fragment groups those MOs that are defined in [2] and model ATM cross-connection facilities and resources.

9. The managed element fragment is built from MOs defined in ITU-T and ETSI recommendations which represent generic telecommunications equipment. These MOs are used for inheritance purposes.

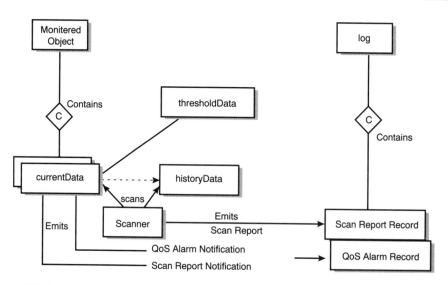

Figure 3.10.5
Entity-relationship diagram for performance model [13]

10. The AN specific managed element fragment. This fragment includes MOs representing AN managed network elements such as OLT and ONU. They have been specifically defined for the access network.

11. The performance collection fragment. The performance part in the AN information model is based on the performance model defined in [13,20,21], draft recommendations which are shown in Figure 3.10.5. In such recommendations several generic MOs for the purpose of performance analysis are defined. For the AN information model only the generic MOs, including the difference scanner defined in the amendment to [20], will be used. It is also possible to use the ATM specific performance objects defined by ETSI [2].

12. The AN specific performance collection fragment. This fragment includes MOs specifically defined for the purpose of performance in the access network. These objects are the specialisation of the performance MOs defined in the previous fragment.

13. The performance monitoring fragment. It includes the MOs that represent VP based performance monitors. An ETSI recommendation [2] defines the performance monitor managed object class.

14. The fault management fragment. In the access network the fault management fragment should include the MOs which represent the two fault management mechanisms defined by ITU-T: continuity check flows and loopbacks (see Figure 3.10.6). The former is a method that allows monitoring of any VPC. This

task is performed by bidirectionalPerformanceMonitor objects linked to VPC termination points. When the source VPC end-point does not send a user cell for a period of time t and no VPC failure is indicated, a continuity check cell is sent. If the sink VPC end-point does not receive any cell within a time interval $t'(t' > t)$, it will send a VP far-end remote failure (VP-FERF) signal to the far end. This mechanism can also be applied to test continuity across any of the VPC segments defined in Figure 3.10.6.

Loopbacks are also intended for fault management. This ATM layer capability allows for OAM cells to be inserted at one location along a virtual path connection and returned (or looped back) at a different location, without having to take the VPC out of service, as recommended by Appendix I of [4]. This capability is performed by non-intrusively inserting a loopback OAM cell at an accessible point along the virtual connection (i.e. at an end-point

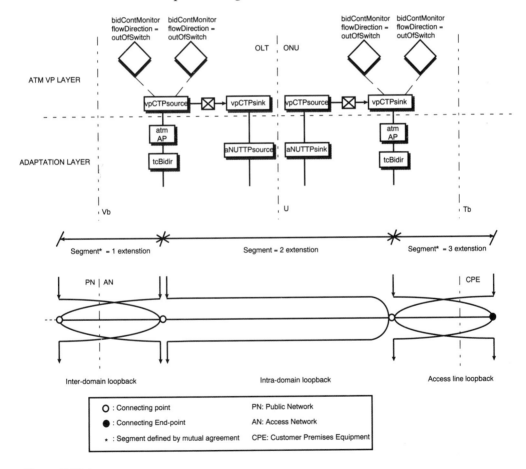

Figure 3.10.6
Types of loopback

or any connecting point). This cell is looped back at any accessible point following an instruction from the OS. A means of confirming that the loopback is performed at the ATM layer, rather than at the physical layer, is provided by requiring the loopback point to change a field within the loopback cell payload. This requirement also overcomes the problem of infinite loopback that could otherwise occur. The applications suggested for the access network are depicted in Figure 3.10.6. These are an intra-domain loopback, an inter-domain loopback initiated by either of both network OSs, and an access line loopback initiated either by the access network or by any of its customers. These loopback flows may also be provided by the bidirectionalContinuityMonitor objects linked to VPC termination points since these objects can easily generate test flows.

The different capabilities to be managed in the NEs are modelled as a set of managed objects included in the previous fragments. Figure 3.10.7 shows the entity-relationship diagram model for the OLT. In this diagram the basic flow of data is depicted. As can be seen, in the OLT the SDH flow is received and terminated (modelled by the 'optical SPITTP' and 'vc4TTP' MOs respectively) and the incoming VP is cross-connected to an internal VP (represented by vpCTP class) and finally transmitted to the ONU/OLT NEs using the AN specific frame relay; the beginning of the new trail is represented by the oNUTTP MO class.

The model of the remote ONUs is shown in Figure 3.10.8. The signal undergoes a treatment similar to that described above.. The ONU receives the AN internal trail and recovers the internal VP. Afterwards, the internal VP is cross-connected to the original VP and presented at

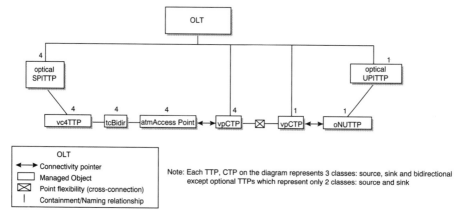

Figure 3.10.7
OLT, connection and containment diagram

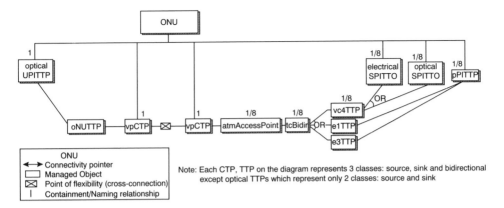

Figure 3.10.8
ONU, connection and containment diagram

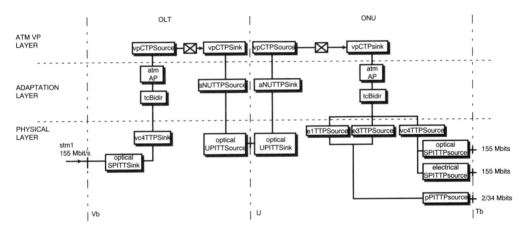

Figure 3.10.9
Layered AN model (downstream)

the Tb interface in SDH format. In Figure 3.10.9 the complete layered model for the AN is depicted. In this diagram the different processes applied to the data flow coming from the B-ISDN to the customer premises equipment can be seen.

Finally, the naming relationships defined in the information model, using the name binding templates, are shown in Figure 3.10.10 for the MO classes contained in the OLT equipment and in Figure 3.10.11 for the ONU MOs. In those diagrams the reader will find the naming, and containment relationships defined in the information model.

3.10.3.3 TMN physical architecture

The TMN functions defined in Chapter 3.10.3.1 may be implemented in a wide variety of physical configurations. Although, in ITU-T

276

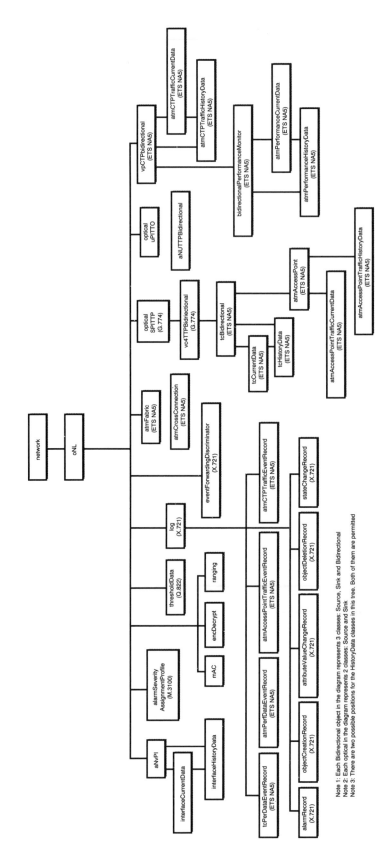

Figure 3.10.10
Naming tree of OLT branch

Note 1: Each Bidirectional object in the diagram represents 3 classes: Source, Sink and Bidirectional
Note 2: Each optical in the diagram represents 2 classes: Source and Sink
Note 3: There are two possible positions for the HistoryData classes in this tree. Both of them are permitted

277

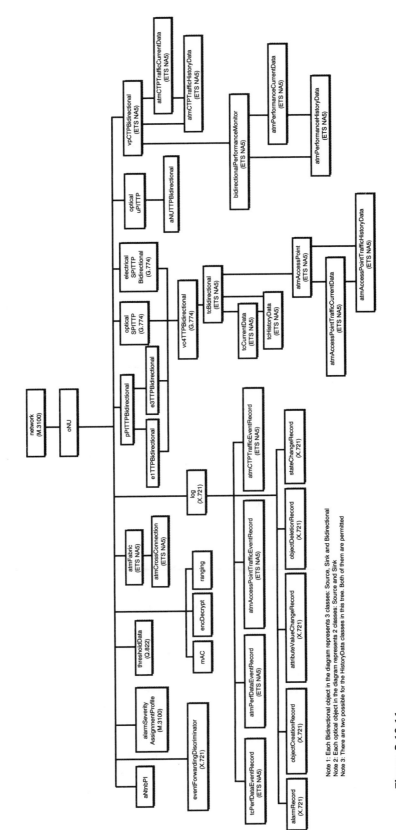

Figure 3.10.11
Naming tree oNU branch

Note 1: Each Bidirectional object in the diagram represents 3 classes: Source, Sink and Bidirectional
Note 2: Each optical object in the diagram represents 2 classes: Source and Sink
Note 3: There are two possible for the HistoryData classes in this tree. Both of them are permitted

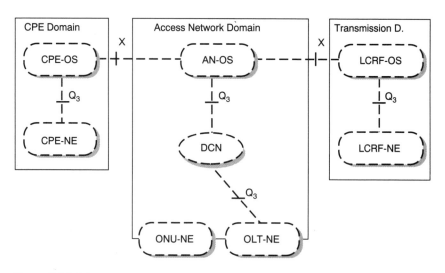

Figure 3.10.12
Physical architecture

M.3010 the relationships between the TMN functional components and the physical blocks have been agreed, such a recommendation does not define the implementation of each physical block. Figure 3.10.12 shows an example of a physical management architecture for the AN using TMN physical blocks. In this case and using the mapping flexibility provided by ITU-T in [7], AN physical blocks have a one-to-one relationship with the functional components envisaged in the functional architecture. In the same way, the reference points between physical blocks become standard interfaces. Figure 3.10.12 represents the management plane of the generic AN architecture depicted in Figure 3.10.2.

OS functions could be implemented in separate OSs with a Q3 interface connecting them over the data communication network. The data communication network (DCN) represents an implementation of the OSI layers 1 to 3. Using the DCN concept, one AN-OS could manage many AN systems. In this case all the OLT network elements belonging to different AN networks will be connected to the same DCN at standard Q3 level. A DCN can be implemented using point-to-point circuits, a circuit switched network or a packet switched network.

The transfer of control and management information between the OS and the OLT management agent is performed via its common Q3 interface and the standard CMIP is used. This information does not travel along the access network. However, the case of the ONU is different. There is no physical Q3 interface between the OS and the ONU. So their control and management information must be routed via the OLT. The access network offers two control channels to achieve these information exchanges: a fast and a slow OAM (operation and maintenance) channel.

An example of time-critical control information may be an order suspending the transmission of an ONU whose signal has been time-shifted and is overlapping other ONU signals. In the access network, the insertion of this kind of information is represented by the monitors linked to the oNUTTPsource and oNUTTPsink objects. These monitors control a so-called segment. The information is transferred to/from the OS from/to the OLT agent via the Q3 interface.

Non-time-critical OAM and management functions like the static allocation of VPI values, bandwidth allocation to virtual path connections, non-critical alarms, encryption keys etc, use the slow OAM channel implemented by means of ATM OAM cells inserted/extracted into/from the user ATM flow. This insertion/extraction is carried out by the monitor linked to the vpCTP source/vpCTP sink object. This information is also transferred between the OS and the OLT agent via the Q3 interface.

The Q3 protocol stack is defined by the ITU-T in Q.811 [22] and in Q.812 [23], but [22], regarding lower layers, does not include the ATM over SDH scenario.

The X interface is, right now, being studied inside the ITU-T groups, and there is no stable recommendation for this interface available. It is assumed that it will be similar to the Q3 interface but with additional functions such as security functions and more communication functions like bulk data transfer (FTAM), electronic mail [14] and EDI capabilities and some directory facilities [15]. In Figure 3.10.13 a proposal of the upper layer protocol profiles for the X interface is shown.

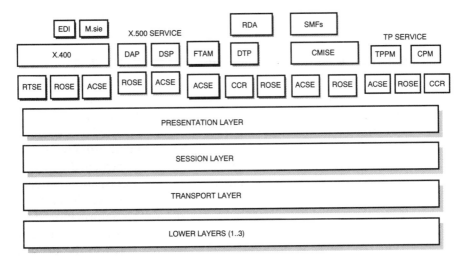

Figure 3.10.13
X interface protocol stack (upper layers)

REFERENCES

[1] RACE CFS H552 Issue E *Managed Objects for Access Networks,* 1994
[2] ETSI DE/NA5-2210 *Version 03 Helsinki 25-29 April 94: B-ISDN Management Architecture and Management Information Model for the ATM crossconnect,* 1994
[3] ITU-T Recommendation G.774 *"Synchronous Digital Hierarchy (SDH). Management Information Model for the Network Element view,*
[4] ITU-T Recommendation I.610 *Integrated Services Digital Network (ISDN) Maintenance Principles. B-ISDN Operation and Maintenance Principles and Functions,*
[5] ITU-T Recommendation M.3010 *Principles for a Telecommunications Management Network,* 1992
[6] ITU-T Recommendation M.3020 *TMN interface specification interface methodology,* 1992
[7] ITU-T recommendation M.3100 *Generic Network Information Model,* 1992
[8] ITU-T recommendation M.60 *Maintenance terminology and definitions,*
[9] RACE I (1024) NETMAN, Deliverable 10 *Telecommunications Management Specifications,*
[10] ETSI prETS 300 304: *Final Draft July 94: Transmission and Multiplexing (TM); Synchronous Digital Hierarchy (SDH) Information Model for the Network Element (NE) view,* 1994
[11] ETSI prETS 300 371: *Final Draft July 94: Transmission and Multiplexing (TM); Plesiochronous Digital Hierarchy (PDH) Information Model for the Network Element (NE) view,* 1994
[12] RACE II (2041) Pan-European Reference Configuration for IBC Services Management, Deliverable 2 *Service and Network Management: Requirements and Reference Configuration,*
[13] ITU-T Draft recommendation Study Group 11 — Report R 2 *Stage 1, stage 2 and stage 3 description for the Q3 Interface. Performance Management,*
[14] ITU-T Recommendation X.400 *Message Handling System,*
[15] ITU-T Recommendation X.500|519 *Information processing systems — Open Systems Interconnection (OSI) — The Directory,*
[16] ISO/IEC 10040 Recommendation *Information Technology — OSI — Systems Management Overview,*
[17] ISO/IEC 10165-1 Recommendation *OSI Structure of Management Information: Management Information Model,*
[18] ISO/IEC 10165-2 Recommendation *OSI Structure of Management Information: Definition of Management Information,*
[19] ISO/IEC 10165-4 Recommendation *Guidelines for the Definition of Managed Objects (GDMO),*
[20] (Draft) ISO/IEC DIS 10164-13 *Information Technology — Open Systems Interconnection — Systems Management — Part 13: Summarisation Function,*
[21] ISO/IEC 10164-11 Recommendation *Information Technology — Open Systems Interconnection — Systems Management — Part 11: Workload Monitoring Function,*
[22] ITU-T Recommendation Q.811 — Lower Layer Protocol profiles for the Q3 interface. March 1992.
[23] ITU-T Recommendation Q.812 — Upper Layer Protocol profiles for the Q3 interface. March 1992.

3.11 FIELD TRIALS EMBEDDING THE BAF SYSTEM

3.11.1 Introduction

Two field trials embedding the BAF system will take place at Turin (Italy), where the system will be connected to an existing ATM test bed located at CSELT (Centro Studi e Laboratori Telecomunicazioni S.p.A., the research centre of the Italian network operator) premises, and at Madrid (Spain), where the system will be connected to the RECIBA network located at Telefónica I+D (the research centre of the Spanish network operator) premises. The Turin trial will be based on an existing PON infrastructure that has previously been deployed for a TPON trial, whereas in Madrid, a new PON infrastructure will be deployed.

The two field trials described in this section, based on the integration of the BAF system into two existing ATM test beds, will allow verification of the operation of the BAF system when loaded with real traffic flows. Experiments will be run on both the physical and ATM layers in order to validate the technical viability of ATM-PON networks, to assess their inter-operability with existing ATM networks, and to evaluate the end-to-end performance of such access systems. More precisely, the objectives are: to demonstrate the inter-operation of the BAF access network with currently available equipment connected to the BAF external interfaces (SDH, 155 Mbit/s optical, and G.703, unstructured 2 Mbit/s), to demonstrate the suitability of the BAF access network for the interconnection of customer equipment and core network equipment, and to demonstrate the transparency of the BAF access network, both from the application and core network points of view.

In the following, a description of the two existing ATM test beds is given, followed by a presentation of the field trial configurations.

3.11.2 The CSELT ATM test bed

The CSELT ATM test bed is the result of the integration of different network elements, realised both by CSELT and other national manufacturing industries, operating companies, public research centres and university laboratories. This project has been partially supported by the National Research Council (CNR) under the auspices of an Italian telecommunication project.

The ATM test bed is located in a show room conceived as an open laboratory for integrating advanced telecommunications systems (innovative equipment, software platforms and end-user applications) used to assess the new services capabilities and features offered by an integrated broadband network. The goal is to provide an environment where provision of broadband services can be emulated

Access to B-ISDN via Passive Optical Networks. Edited by U. Killat
© 1996 John Wiley & Sons Ltd

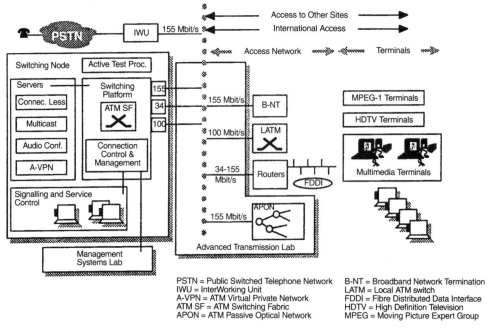

Figure 3.11.1
Configuration of the CSELT ATM test bed

and demonstrated in a laboratory, providing useful information for field deployment. The test bed includes various types of equipment, falling into different categories, terminals, access network and switching node, as depicted in Figure 3.11.1.

3.11.2.1 Terminals

Terminals, located in the customer area, are based on general purpose machines that, by appropriate I/O peripherals, have functionalities to satisfy the requirements of distributed multimedia applications (video communication, multimedia file transfer, access to remote databases, etc.). The following terminals are available in the test bed.

- Multimedia terminals with ATM interface, supporting isochronous constant bit-rate (CBR) traffic for audio, video and connectionless variable bit-rate (VBR) traffic for data;

- ATM terminals, based on UNIX workstations and commercial ATM boards, supporting VBR traffic for data;

- Digital HDTV set (decoder and monitor) with ATM interface;

- Audio conference terminals, to be used in conjunction with the audio conference server;

- MPEG-1 terminals for MPEG-1 coded audio-visual sequence retrieval;
- High-definition or MPEG-encoded digital video sources.

3.11.2.2 *Access network*

The access network is located between the customer area and the public network operator (PNO) area. The equipment belonging to this category includes:

- Ethernet and FDDI routers, performing IP datagram routing via the ATM network;
- local area networks based on ATM technology, using commercial local ATM switches;
- Broadband network terminations, multiplexing ATM streams from the terminals onto the network access and distributing the network ATM streams to the terminals;
- ATM PON as the access medium to the switching node area;
- SDH (synchronous digital hierarchy) transmission system.

3.11.2.3 *Switching node*

The switching node is located in the PNO area, with parts (servers) that can reside in the service provider area. The switching node structure is the result of the integration of several blocks.

- Switching platform, with the following subsystems:
  - (a) The ATM switching fabric, offering capabilities for handling VP and VC connections on the following line interfaces at UNI and NNI: 34 Mbit/s PDH, 155 Mbit/s SDH, and 100 Mbit/s ATM Forum. The switching fabric performs label and multiplex switching of ATM cells at connection level, supporting multicasting of ATM cells by means of a multicast server (see below), usage parameter control (UPC) and network parameter control (NPC).
  - (b) The connection control and management (CC&M) system is a distributed environment supporting telecommunications services in the connection handling within the switch. The CC&M system manages and controls connections and network resources (VP and VC identifiers, bandwidth, paths, etc.). The CC&M provides knowledge of the mapping between logical entities and physical objects, knowledge of network topology for the purposes of routing and connection set-up, modification and clearing.

- Servers: The switching platform is equipped with a set of elements (servers) that offer special services:

  (a)   The connectionless server (CLS) supports a broadband connectionless data service on an ATM network. The connectionless server is a specialised service provider able to handle a connectionless protocol and to realise the adaptation of the connectionless data units into ATM cells to be transferred in a connection oriented environment. The connectionless protocol includes functions such as routing, addressing and quality of service selection.

  (b)   The audio conference server offers an audio conference service to terminals, using access signalling. The server manages and controls the conferences; it allows for the adding and dropping of parties to the conference. Moreover the server performs operations on the voice samples coming from the different participants and uses the multicasting capabilities of the switching platform to distribute the result.

  (c)   The multicast server is able to handle point-to-multipoint connections. When a client (e.g. the audio conference server) requires the establishment of a multicast connection, the server defines the duplication and reassembling points for the streams inside the switching node. It extends the TDI functions, offering primitives to create and delete multicast connections, and to add or remove single connection branches.

  (d)   The A-VPN (ATM — virtual private network) Server offers a rich set of functionalities, like selective cell discarding at message level for congestion control, large buffering capacities, traffic shaping for congestion avoidance and statistical multiplexing purposes, handling of different priority service classes, and fast resource management functions.

- signalling and service control; this part is responsible for the control of the telecommunications services to be provided to the end users, by exploiting capabilities of the underlying network and by communicating with peer entities in remote locations.

The test bed includes also a management systems laboratory for the evaluation of management systems. This laboratory is based on an open infrastructure in order to be interconnected to different network environments within CSELT and outside. The main focus of the activity of this laboratory is on the ATM and SDH management systems for both the public and the private networks. The CSELT ATM test bed is part of a national ATM pilot network that includes an international gateway towards the PEAN located in Milan and three national nodes located in Turin, Rome and Milan.

3.11.2.4 Services and applications

The following services are currently supported by the test-bed.

- Bearer and network services
    - Permanent Virtual Paths (PVPs),
    - Permanent and switched virtual channels (PVCs and SVCs),
    - Set-up of switched services by signalling;
- Telephony, between multimedia terminals and PSTN telephones, using AAL-1 and 64 kbit/s SVCs;
- Broadband videotelephony, between multimedia terminals, providing H.261 [1] or H.320 [2] video conference services, using AAL-1, a 64 Kbit/s to 2 Mbit/s SVC for video and a 3–7 kHz bandwidth SVC for audio;
- Audio conferencing between audio conference terminals and the audio conference server, using access signalling;
- IP datagram routing, between multimedia terminals, routers and the connectionless server, using AAL3/4 or AAL5 and a 10 Mbit/s PVC or SVC;

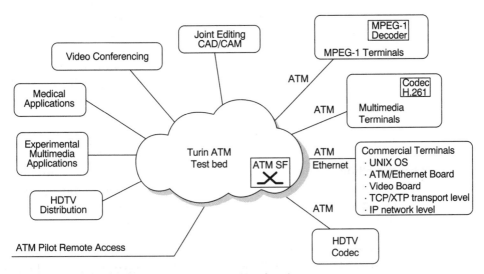

CAD/CAM = Computer Aided Design/Computer Aided Manufacturing
ATM SF = ATM Switching Fabric
HDTV = High Definition TeleVision
MPEG = Moving Picture Expert Group
UNIX OS = Unix Operating System
TCP/XTP = Transmission Control Protocol/eXpress Transfer Protocol
IP = Internet Protocol

Figure 3.11.2
Applications and terminals in the CSELT ATM test bed

- Digital HDTV delivery, from the HDTV source in the service provider area to the HDTV Terminal in the customer area, using AAL-1 and a 45–100 Mbit/s PVC.

The applications currently demonstrated on the test bed are (Figure 3.11.2):

- Cooperative work (video conferencing and joint editing) between multimedia terminals, including medical applications;

- LAN interconnection across an extended geographical area, running client–server applications;

- Interactive application for multimedia information retrieval using real-time MPEG-1 decoding for displaying live video on a 1.5 Mbit/s SVC;

- B-ISDN/PSTN interworking, by means of a telephone call between a multimedia terminal and a PSTN telephone;

- Video on demand service, based on MPEG-encoded video at 1.5 Mbit/s;

- HDTV signal distribution.

3.11.3 The Telefónica ATM test bed

The experimental B-ISDN network RECIBA was the first integrated Spanish project in the pre-commercial broadband field. RECIBA, defined, owned and sponsored by Telefónica, and developed by Telefónica I+D, is located in Madrid and has been available since the end of 1992.

The main objectives of the RECIBA project are:

- Availability of an experimental network on which it is possible to test different aspects of installation, operation and maintenance of broadband equipment (network, terminals, service provider centres, etc.), anticipating the first commercial systems;

- A means of technical and economical evaluation of the impact of the international recommendations emerging nowadays in the broadband field;

- Empirical measurement and evaluation of different quality of service (QoS) parameters (network propagation delay, cell loss probability, subjective audio and video quality, etc), thus helping to specify the mandatory characteristics for future network equipment and broadband services both in their individual behaviour and in their joint exploitation;

- Implementation of a national broadband demonstrator as the result of expertise and effort formerly dedicated exclusively to European cooperation projects (mainly RACE);
- The use of RECIBA as an open platform for integrating all types of broadband applications, thus enabling the development of new equipment, interfaces and functions, as well as the participation and collaboration of institutions other than Telefónica.

Figure 3.11.3 shows the broadband network concept underlying the RECIBA project. In order to trace the parallels between the general structure of this experimental network and that of other conventional ones, the following breakdown of the network may be defined:

- The customer premises network including the network terminations NT2, the T_B and S_B interfaces, the terminals and the service provider centres (SPC) connected to the NT2 network elements,
- The access network including the U_B interfaces and the broadband network terminations NT1, and the
- The transport network including the ATM VP Cross-connects (VPCC) and the network to network interfaces I_R connecting them.

3.11.3.1 *Customer premises network*

- The broadband NT2 is the key element of the customer premises network. This network element is able to establish ATM connections on both a virtual path and virtual channel basis. Local or remote establishment of 2 Mbit/s connections by means of VPCs is handled through operator commands on the man–machine interface terminal, because of the lack of signalling for interworking between narrowband and broadband networks. Both local and remote operation of the NT2 is supported. The terminals are based on general purpose machines, with the peripherals and terminal adapters necessary to comply with the requirements of the supported broadband services and applications. The following terminals are available in the RECIBA network:
    - Multimedia terminals developed within different RACE and internal projects, which support any kind of multimedia (audio, video and data) application,
    - Videophone terminals,
    - Workstations with ATM interfaces,
    - HDTV professional and consumer monitors,
    - Terminal adapters for interworking between Ethernet, FDDI, and G.703 and ATM,
    - Service provider centres for multimedia applications.

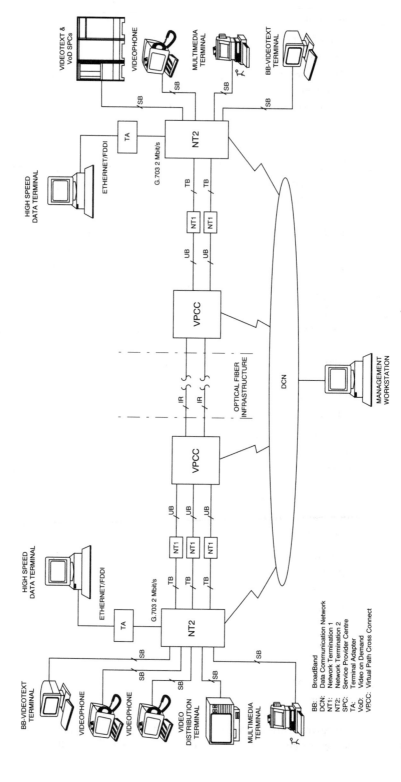

Figure 3.11.3
Configuration of the RECIBA ATM test bed

3.11.3.2 Access network

This consists basically of the broadband NT1 network element, implementing the interface between the public and the private network domains. This network element behaves as an optical regeneration device that also implements monitoring and loop functions in order to facilitate the operation, administration and maintenance of the network.

3.11.3.3 Transport network

The VPCC network element is the heart of the transport network in RECIBA. It behaves as an ATM VP cross-connect, providing semi-permanent and flexible virtual path connections between any of its interfaces. In the same way as for the NT2, local or remote establishment of (G.703) 2 Mbit/s connections by means of VPCs is handled through operator commands on the man–machine interface terminal. The VPCC architecture is composed of the following subsystems:

- **Switching subsystem** This subsystem is the VPCC central switching kernel. Different static topologies are supported when configuring the switching fabric, depending on the throughput needed. Each node is implemented by means of a 4×4 640 Mbit/s switching matrix. Three out of those four links will be used to connect the switching node to its neighbouring nodes in the switching fabric. The fourth one will act as a traffic concentrator for the distribution and MUX subsystem. All links exchange extended 56-octet ATM cells.

- **Distribution and MUX subsystem** This subsystem is built around different interface termination elements, which implement the U_B, I_R, and (G.703) 2 Mbit/s interfaces. Up to eight interface termination elements are supported by one switching node. Two U_B, or I_R (SDH, 155 Mbit/s,) optical and four 2 Mbit/s interfaces are implemented on each board.

- **Control subsystem** This subsystem implements the relevant control and OAM functions of the network element and interfaces to the management system. Both local and remote operation of the VPCC are supported.

3.11.3.4 Management system

The RECIBA network includes also a management system, which follows a TMN architecture, supporting the following functional management areas:

- Configuration management includes a set of functions identifying the resources to be managed, enabling the installation or closing down of both hardware and software components within a distributed system, and performing the allocation of software components to hardware components. Configuration management provides also all the necessary information for the proper up-keep of the connection tables, assigning VPI values to the connections and contributing to the call acceptance procedures.

- Performance management monitors the quality of service provided to virtual path connections, supports traffic statistics at each interface between network elements, and implements the bandwidth allocation and policing functions on a per-VP basis.

- Fault management provides alarm/failure detection, testing, isolation and recovery measures for the different components, including the status of external and internal interfaces. This area communicates faults to other management areas to report several changes in the network (configuration and security), and get several measurements regarding error rates (performance).

- Security management ensures secure access to management information by authorised management information users both horizontally (limiting the access to defined management domains in the network), and vertically (limiting the access to subsets of management functions).

The management system supports local management of each network element or remote management through a data communication network (Ethernet or X.25).

3.11.3.5 *Applications*

The following services and applications are currently supported by the RECIBA network.

- **Broadband videotext services** In RECIBA there are different applications based on this type of service i.e. a hotel booking application and a teleshopping application. These applications use a workstation with suitable peripherals as hardware platform, and terminal adapters for interworking between Ethernet and ATM. The service provider centres contain multimedia databases with text, graphics, high resolution fixed image video, video sequences and high-fidelity audio.

- **High quality videotelephone service** This is an audiovisual bidirectional and symmetrical service, allowing real-time communication between two users with high quality sound and moving image exchange. The basic terminal features are: local image view, local image switching between two cameras, local image transmission suppression, icon-based user information, interworking with standard telephones and hands-free facility.

- **Video message service** This service uses the same basic terminal as for the high quality videotelephone service. The functions are very similar to the ones provided by a conventional audio message service.

- **HDTV distribution service** The signal is distributed from a TV service provider centre with two associated audio channels. The application provides the following facilities: audio channel switching, image freezing, channel exploring, mosaic presentation of all channels, iconised information and user help.

- **Video on demand service** This service is an information retrieval application enabling user access to long duration and high quality audiovisual documents, i.e. movies and video clips, over an HDTV home terminal. The service provider centre contains multimedia databases with text, graphics, high resolution fixed image video, 1.5 Mbit/s MPEG-1 encoded video sequences, and high-fidelity audio. The application supports a navigation menu enabling access to multimedia information about the titles available, like a trailer of the movies, biography of players, actors, directors, etc. The application provides all the features of a home video. It is also possible to switch between the video on demand service and the HDTV distribution service at any time. It is even possible to have a dual screen with both applications.

- **Multimedia multipoint videoconference and cooperative work services** In RECIBA there are three different applications based on those services. Two of them use a workstation with suitable peripherals as a hardware platform, and terminal adapters for interworking between Ethernet and ATM. The other one uses a multimedia terminal specifically developed for the application. All of them support videoconferencing, common manipulation of texts, white board, data files interchange, distributed pointer, etc.

- **LAN interconnection service** Both, LAN interconnection through ATM terminal adapters and ATM LANs are services available in RECIBA, and are used in some of the applications mentioned above.

3.11.4 The BAF field trial in Turin

The scenario for integration of the BAF system into the CSELT ATM test bed is shown in Figure 3.11.4. The BAF system will act as the

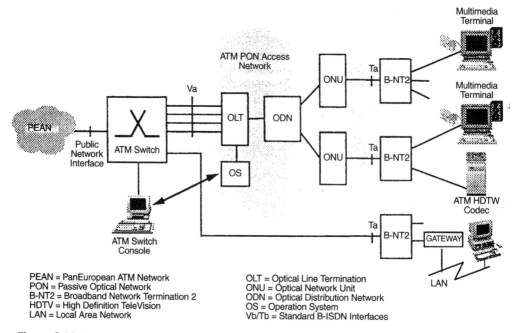

Figure 3.11.4
Turin field trial scenario

access network between the ATM node on one side and the customer premises networks on the other side. The connection of the ATM node to the pan-European network is also shown.

The integration of the BAF system into the CSELT ATM test bed will require the adaptation of the B-NT2 and ATM node physical interfaces to the T_B and V_B interfaces of the BAF system. VP switching, performed by the ATM node, will be adapted to the BAF system requirements. Local signalling between the ATM node and the terminals will be adapted if needed. The existing PON network will be extended, where necessary, to reach the BAF terminals. No direct interconnection will be established between the ATM node console and the BAF system. This interaction will indeed be based on human operator intervention. Further developments will possibly include the interconnection of the BAF management system and the management system of the ATM node via a standardised management interface (X interface).

The aims of the field trial are twofold: first it will allow assessment of the service and network performance throughout the ATM-PON access network by means of experiments in the ATM adaptation layer, in the ATM Layer and in the physical layer. These experiments will be carried out using the broadband applications made available in the ATM test bed. Secondly, the field trial will act as a comprehensive demonstration environment for the ATM-PON concept.

3.11.5 The BAF field trial in Madrid

The RECIBA network constitutes the kernel of the Spanish National Host, which means that this network will be used by ACTS and other projects in the near future, as a platform for experimenting with broadband services and applications, most of them involving real users, in order to assess the commercial possibilities of such new services. The RECIBA test bed is also connected to Telefónica's ATM pilot network and, through it, to the pan-European ATM Network, assuring in this way connectivity to other broadband islands. Therefore, it will be possible for the limited field trial presented in this section, performed within the BAF project, to evolve into a more ambitious field trial, involving different locations and more applications.

The proposed field trial is based on interconnecting the BAF access system with the RECIBA network, by substituting one of the U_B interfaces of that network with the access system. The field trial configuration is shown in Figure 3.11.5. In this configuration standard interfaces between the access system and the RECIBA network and between the terminals/service provider centres and the access network will be used. The RECIBA and BAF management systems will be independent, which means that no direct interconnection will be established between the RECIBA management system and the BAF system.

The RECIBA ATM network will have the functionality of an ATM VP cross-connect, providing semi-permanent connections through the BAF access system and the service provider centres between the terminals at the customer premises.

Local applications will be used in order to load the BAF demonstrator with different types of traffic. This will allow verification of how the access network is able to transport different types of traffic while providing the required quality of service. The obtained performance measurements, basically in terms of QoS parameters, will be used to validate the implemented functions and the conformance with international requirements.

The subset of services and applications available in RECIBA and supported by the BAF field trial is described in more detail in the following paragraphs.

- Computer supported cooperative work (CSCW), distance learning and real-time multimedia multipoint videoconference applications were developed within the ISABEL/IBER RACE II projects. These applications use a multimedia terminal based on a Sun-10 workstation running the SunOS 4.1.3 operating system and X windows version 11.5, with suitable interfaces such as video micro-camera, Parallax board with JPEG videocodecs, a speaker

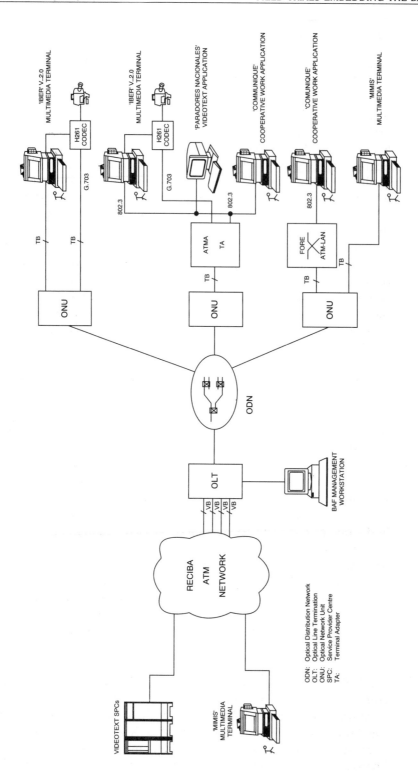

Figure 3.11.5
Madrid field trial scenario

box, and a microphone. It is also possible to extend the configuration for the distance learning application with a video projector, a large projection screen, H.261 videocodecs, echo cancellers, a set of high quality microphones, stereo speaker boxes, amplifiers and professional video cameras, in order to improve the quality of the service in virtual conference room environments. The applications make use of different tools such as a video distribution with live video support, audio distribution, cooperative editor, shared white board, slide presentation and pointer/pencil. The high quality distance learning service uses a (G.703) 2 Mbit/s interface, whilst the rest of applications supported may use Ethernet or ATM interfaces. The total bandwidth required by the applications depends on the number of users, the quality factor used, the speed and the window size. The video quality and the speed may be adjusted by the application or by the user at any time, as well as being remotely controlled, in order to comply at any time with the bandwidth available. Of course, if an Ethernet interface is used, the bandwidth available is less than for an ATM interface. Two different configurations are used for the multimedia terminal. In the first configuration the multimedia terminal is connected to a terminal adapter available in the RECIBA test-bed. This adapter provides a 2 Mbit/s (G.703) and an Ethernet interface. In the second configuration the multimedia terminal makes use of the T_B and T_{NB} interfaces of an ONU of the BAF system. In this way, in particular the narrowband board developed within the BAF project is tested for a different implementation.

- "Communiqué" is a commercial CSCW and virtual conference application, which uses a multimedia terminal based on a workstation running the OpenLook window manager, with suitable interfaces such as video micro-camera, a video board, a speaker box, and a microphone.The application makes use of different tools such as video control, an audio control, a text tool for distributing text to conference members, a shared white board tool, a cooperative editor, and a file distribution tool. The connection of one of the workstations to the ATM network is made through a terminal adapter developed outside the BAF project. The connection of the other workstation is made through commercial ATM-LAN equipment.

- A multimedia interactive interpersonal conferencing system is available as a result of the RACE II MIMIS project. MIMIS comprises CSCW and video conference applications, which use a multimedia terminal based on a SUN workstation running the SunOS 4.1.3 operating system and X windows version 11.5, with suitable interfaces such as video micro-camera, two video boards, a speaker box, a microphone and a smart-card reader for authentication purposes, plus a terminal adapter specifically developed for MIMIS. The applications support different conference facilities such as high quality audio (G.722 for conversational and MPEG-1 audio for retrieval service), motion video (H.261 for conversational and MPEG-1 video for retrieval service), still images in JPEG format with pointer and marker, file transfer, exchange of messages, joint viewing, joint editing, multimedia document handling and security functions including authentication, encryption at ATM layer, integrity and confidentiality.

- The trial configuration also includes a multimedia videotext application called "Paradores de Turismo", completely developed by Telefónica, which provides access from multimedia terminals to a service provider centre containing multimedia information (video, still images, audio and data) on a Spanish hotel network. The multimedia terminal is based on a workstation running OpenLook window manager and equipped with suitable interfaces such as a video board and a speaker box. The connection of the terminal to the ATM network is made through a terminal adapter. The service provider centre is also based on a Sun workstation with interBase database, proper storage devices, and an ATM interface card from Fore. The application makes use of different tools such as a navigation menu able to organise the information by geographical location, services available and price criteria. It provides multimedia information about each hotel (a video trailer of the facilities and location of the hotel, including audio information) and textual information about the surroundings, services available, history of the building and prices, and real-time booking possibilities.

3.11.6 Conclusion

The field trials described will allow assessment of the BAF system performance in a realistic environment, paving the way for more extended trials involving real users. Indeed, plans for the near future include the incorporation of the BAF access network into the Italian National Host (an experimental platform made available by the Italian Government to host the experiments of projects carried out within the framework of EU programmes and other international and national R&D programmes) and the provision of broadband multimedia services to both professional and residential users.

REFERENCES

[1]   ITU-T Reccommendation H.261 *Video codec for audiovisual services at p ×*
       64 kbit/s
[2]   ITU-T Reccommendation H.320 *Narrow-band visual telephone systems and*
       terminal equipment

Index